T0259717

SpringerBriefs in Physics

More information about this series at http://www.springer.com/series/8902

Vladimir Pletser

Gravity, Weight and Their Absence

 Springer

Vladimir Pletser
Technology and Engineering Centre
  for Space Utilization
Chinese Academy of Sciences
Beijing
China

ISSN 2191-5423   ISSN 2191-5431   (electronic)
Springerbriefs in Physics
ISBN 978-981-10-8695-3   ISBN 978-981-10-8696-0   (eBook)
https://doi.org/10.1007/978-981-10-8696-0

Library of Congress Control Number: 2018934907

This Springer imprint is published by the registered company Springer Nature Singapore Pte Ltd.
part of Springer Nature
The registered company address is: 152 Beach Road, #21-01/04 Gateway East, Singapore 189721,
Singapore

Preface

Hello!

My name is Vladimir Pletser. Space and astronautics have been a passion since childhood. Understandably, my studies and my career were oriented towards space research. I studied mechanical engineering, then geophysics and I finished with a Ph.D. in physics at the Catholic University of Louvain, in Louvain-la-Neuve, Belgium.

I still remember one of my professors explaining what was weightlessness and the enlightenment that I felt of understanding something that sounded complicated but that eventually was so simple. Weightlessness is obtained when there is free fall, and vice versa, and nothing else!

In the mid-80s, I joined the European Space Research and Technology Centre of the European Space Agency, located in Noordwijk in the Netherlands. During the 30 years, I spent at the Agency, I was in charge among other projects of the aircraft parabolic flight programme. During these flights, we would recreate weightless conditions for approximately 20 s during parabolas that the aircraft was performing in the air. And this notion of weightlessness was always there with me.

In this book, I would like to introduce you to the wonders of weightlessness. Can we live without weight? Or is it without gravity? I hope that the following pages will help you to better understand some simple concepts, although these are not always well understood even by eminent scientists.

The book is arranged in four chapters. Chapter 1 introduces the basic notions of inertia, inertial and non-inertial reference frames, gravitation, gravity and weightiness. Chapter 2 defines the concepts of weightlessness and free fall. Chapter 3 addresses microgravity, the methods to create it, its importance and applications, in particular for material and fluid sciences. Finally, Chapter 4 deals with the effects of weightlessness on human physiology and the changes the body of astronauts will go through during a spaceflight.

In the hope of creating a text that flows, I have reduced the number of formulas in the main text leaving all the detailed developments in the appendices.

I hope that I have been able to share my passion and wish you an enjoyable read. If you have any questions, do not hesitate to contact me.

Vladimir Pletser

Assistant Professor, Department of Physics, Faculty of Sciences University of Kinshasa, Congo; and Catholic University of Louvain, Louvain-la-Neuve, Belgium (1982–85)

Senior Physicist—Engineer, European Space Research and Technology Centre (ESTEC), European Space Agency (ESA) Noordwijk, The Netherlands (1985–2016)

Visiting Professor—Scientific Adviser, Technology and Engineering Centre for Space Utilization Chinese Academy of Sciences, Beijing, China (since 2016)

Acknowledgements

The author wishes to thank Dr. D. Poelaert (ESA) and Prof. D. Johnson (Catholic University of Louvain, Belgium) for fruitful discussions on inertial reference frames, inertial forces and translation issues in Chapter 1. The author wants to thank Prof. M. Hinsemkamp (Free University of Brussels) for his suggestions during the initial preparation of the text of this Chapter 4.

The support of Dr. Q. Cao (Technology and Engineering Centre for Space Utilization, Chinese Academy of Sciences, Beijing), Dr. D. Pletser (Imperial College, London), Dr. J. Pletser-Dent and Mr. O. Shearman (International Baccalaureate Organization, The Hague) and ESA Astronaut Jean-Francois Clervoy in proofreading the text is highly appreciated.

M. Pierre-Emmanuel Paulis put his talent of draughtsman to produce some of the illustrations of this text (Figs. 1, 2, 7 and 9).

The author is supported by the Chinese Academy of Sciences Visiting Professorship for Senior International Scientists (Grant No. 2016VMA042).

Contents

Notes

A first version of the text of this book was initially published in French in a series of four articles published in *Ciel et Terre*, the Bulletin of the Royal Belgian Society of Astronomy, Meteorology and Geophysics, Brussels, Belgium, in 1997 and 1998, as follows:

Pletser V. "Pesanteur ? Non, merci ! 1ère partie : L'inertie peut-elle nous faire bouger ?", *Ciel et Terre*, Bulletin de la Société Royale Belge d'Astronomie, de Météorologie et de Physique du Globe, **113-1**, 9–17, 1997.
Pletser V. "Pesanteur ? Non, merci ! 2ème partie : L'impesanteur est-elle sans gravité ?", *Ciel et Terre*, Bulletin de la Société Royale Belge d'Astronomie, de Météorologie et de Physique du Globe, **113-2**, 53–61, 1997.
Pletser V. "Pesanteur ? Non, merci ! 3ème partie : Faut-il s'appesantir sur la microgravité ?", *Ciel et Terre*, Bulletin de la Société Royale Belge d'Astronomie, de Météorologie et de Physique du Globe, **113-4**, 153–162, 1997.
Pletser V. "Pesanteur ? Non, merci ! 4ème partie : Effets physiologiques graves de l'impesanteur" (in French), *Ciel et Terre*, Bulletin de la Société Royale Belge d'Astronomie, de Météorologie et de Physique du Globe, **114-3**, 96–103, 1998.

Chapter 1
Can Inertia Make Us Move?

1 Introduction

Gravity disturbs certain experiments and reduces the field of investigation of some scientific domains. *Gravity* is omnipresent on Earth. It accompanies us in all our movements since birth and it is difficult to imagine our environment without it. How to conceive that, without it, the slightest step would propel us in the air or that liquids would take a spherical form?

And still, *gravity* effects hide other effects pertaining to materials or fluids under study and that depend often on intrinsic properties of matter or of its state. Convection in fluids, so evident that it is called "natural", is caused by *gravity* acting on local differences of density caused by differences of temperature or concentration. The resulting Archimedes or buoyancy force induces an ascending motion of fluid zones of lesser density and a descending motion of fluid zones of larger density, creating convection cells in gases ("warm air rises, cold air descends"), in liquids and in melted solids, yielding disruptive phenomena in separation processes.

Nevertheless, *gravity* can only exist if the gravitational attraction is thwarted by the presence of a solid surface that prevent the gravitational force of continuing its action of attraction. Although gravitation is omnipresent in the universe, *gravity* is the exception and its absence, weightlessness, the general rule.

This book is subdivided into four chapters. This chapter recalls some notions in physics on inertia, inertial and non-inertial reference frames, gravitation and *gravity*, that are needed to understand the phenomenon of weightlessness. The second chapter is devoted to the concepts of free fall and weightlessness. Microgravity is the subject of the third chapter with its methods, means and applications. The fourth chapter covers the physiological effects of weightlessness.

This book is not a rigorous lesson of classical mechanics. Certain notions are introduced while relying on the reader's intuition. The well-informed reader will forgive this approach that has the advantage of recalling rapidly the necessary notions to the introduction of the concept of weightlessness. The text is structured such

© The Author(s) 2018
V. Pletser, *Gravity, Weight and Their Absence*, SpringerBriefs in Physics,
https://doi.org/10.1007/978-981-10-8696-0_1

that important notions are introduced in the main part and details and mathematical developments are relegated to appendices whose reading is not strictly necessary.

Let us mention also that the word 'gravity', so far italicized, is used in the English language to designate two different, although close, physical concepts that will be explained in more details in Sect. 3. However, to conform to the usage, we have kept the word 'gravity' everywhere but in italics, except when it means the force associated with the phenomenon of gravitation. The meanings of the word 'gravity' are discussed in Appendix 1.

2   Inertia and Inertial Reference Frames

Among numerous experiments with projectiles in the 17th century, Galileo was interested by the motion of a spherical bead on a horizontal plane. The launched bead rolls along a straight line and eventually stops because of friction. Galileo asked himself what would happen in the theoretical ideal case where the bead would be perfectly spherical, the plane infinite and perfectly horizontal, and friction non-existent. He had the audacity (for the epoch) to affirm that the bead would continue its motion indefinitely along a straight line, at a constant speed and keeping its initial velocity. In the particular case where the initial velocity is null, the bead would stay in its initial rest state. These two visions of things seem to us perfectly logical and natural. This simple experiment introduces Galileo's principle of inertia, the first to have been stated. If no forces are applied to the bead (or if the sum of all applied forces is null), the bead keeps its initial velocity, or in other words, its acceleration (i.e. the rate of change of its velocity) is null. This bead is unable by itself to change its velocity without an external force: this body is inert (from Latin *iners*, unable).

Newton has generalized this Principle of Inertia to all bodies, whatever they are, terrestrial or celestial, in a natural undisturbed motion:

> A body on which no force is acting is either at rest or in a uniform rectilinear motion.

i.e. with a constant velocity and along a straight line. If a body is at rest, it tends to stay at rest; if it is in motion, it tends to stay in motion. This statement is relatively simple, but it was diversely understood through centuries and still creates havoc.

To formulate this inertia principle, we need to recall some general basic notions. First of all, the mass of a body: intuitively, everybody can imagine what it is. However, several definitions are possible and were the object of numerous discussions and debates. We take here the definition given by Newton:

> The mass of a body is the measure of the quantity of matter in this body.

This definition can be mistaken for that of another notion and must be understood correctly (see Appendix 2). Two other definitions of the mass are given further, but for the moment, this one is sufficient to introduce what follows. The magnitude of the mass is expressed by a scalar (i.e. by a single number) and its unit in the standard SI[1] system is the kilogram. Recall as well that mass is different from weight, which will be introduced further on.

Other physical entities, like velocity and acceleration, are expressed by vectors, i.e. by two factors, the norm (or magnitude) and the direction of the vector.

In what follows, we will assume that bodies have constant mass and move as a whole, without deformation, and at velocities sufficiently smaller than the speed of light (approximately 300 000 km/s, or slightly more than a billion km/h) to neglect all relativistic effects.

Physical vectorial entities are always expressed with respect to a reference. To say that a car drives at 100 km/h means that its speed is 100 km/h with respect to the road (whose velocity is null). In Galileo's experiment, the velocity and acceleration of the bead are expressed with respect to the horizontal plane (that can be assumed at rest). One introduces so the notion of reference or referential frame. One distinguishes two kinds of reference frames: the inertial (or Galilean or Newtonian) reference frame and the non-inertial (or non-Galilean or non-Newtonian) reference frame.

> An inertial reference frame is a reference frame in which the inertia principle is verified.
> Any other reference frame is non-inertial, i.e. the inertia principle is not verified in this reference frame.
> A reference frame in a uniform rectilinear motion with respect to another inertial reference frame is also an inertial reference frame.

The inertial reference frame is an ideal theoretical concept (see Appendix 3). However, everybody has experienced the phenomenon of inertia in a non-inertial reference frame during a journey by car, train, bus or plane. What pushes you to go in the opposite direction of the vehicle acceleration is caused by the phenomenon of inertia, defined as follows:

---

[1]International System of Units (in French: *Système international d'unités*, SI).

Inertia is the physical phenomenon coming from the property of a mass to oppose the change of its velocity, in speed (the magnitude of the velocity vector) or in direction.
The inertia force is the resistance that a mass opposes to changes of its velocity.

In the non-inertial reference frame attached to the vehicle, this vectorial force of resistance or of inertia $\vec{F}_{in}$ reads like the product of the mass m of the body and its acceleration $\vec{a}$:

$$\vec{F}_{in} = -m\,\vec{a} \tag{1}$$

and where the minus sign is conventional to indicate that the inertia force is in the opposite direction of the acceleration

What does it mean? Let us take the example of a passenger in a train walking toward the front of the train, moving forward. Let us consider the three following cases, as illustrated in Fig. 1:

(1) As long as the train moves in a straight line at a constant velocity, the passenger (if not looking through the window) can think that he walks without being shaken and that the train may as well be at rest, i.e. that the passenger does not feel any inertia force. One can say that the reference frame of the train is inertial and in rectilinear (i.e. along a straight line) and uniform (i.e. at constant speed) motion, with respect to the ground, that can be considered in a first approximation as an inertial reference frame.
(2) If the train, while keeping its speed constant, turns to the right (or to the left), the walking passenger is thrust to the left (or to the right).
(3) If the train, while still moving in a straight line, starts to accelerate (or to decelerate), the forward walking passenger is thrust backward (or forward).

In these last two cases, the reference frame attached to the train is no longer inertial because it is no longer in a rectilinear motion in case (2) and because it is no longer in a uniform motion (the speed is no longer constant) in case (3). In these two cases, the passenger feels an inertia force that pushes him in the direction where he would tend to go and with the velocity that should be his, would the rectilinear uniform motion not have been thwarted. (see Appendix 4 for more on this).

More generally, any reference frame that is not in a rectilinear uniform motion with respect to an inertial reference frame, is non-inertial. One recalls that:

In an inertial reference frame, there are no inertia forces.
In a non-inertial reference frame, there are always inertia forces.

Which inertial reference frame can be considered in practice? Let us recall first that a spatial reference frame, designated by R, is made of three axes perpendicular

Fig. 1 Example of the passenger walking toward the front of the train. (1) The train drives in a straight line at a constant velocity; its reference frame can be considered as inertial. The passenger does not feel anything. (2) The train turns to the right at constant speed; its reference frame is no longer inertial. The passenger is thrust to the left. (3) The train accelerates in a straight line; its reference frame is no longer inertial. The passenger is thrust backward. (Credit: P. E. Paulis)

to each other, designated by X, Y and Z, coming from a to-be-defined origin. Any three-dimensional vectorial entity is expressed in such a reference frame.

Rigorously speaking, an inertial reference frame must be either at rest or in a uniform rectilinear motion with respect to another reference frame which itself must be inertial. In practice, it is never the case. Nevertheless, two reference frames can be considered as inertial with a good approximation to study phenomena on our planet.

Reference frames attached to Earth have as origin a point of the planet, in particular either the centre of the Earth (the geocentric reference frame attached to the Earth, sometimes called Earth-centred, Earth-fixed) or a point on the surface of the planet (a reference frame attached to Earth's surface, e.g. the reference frame of a laboratory on Earth), and as axes three axes fixed with respect to the Earth considered as a perfect non-deformable solid. These reference frames can be considered as inertial in a first approximation, as long as the phenomenon under study in one of these reference frames is of short duration, of limited spatial extend with respect to Earth's dimensions and unaffected by Earth's rotation. The committed approximation is sufficient to study most of problems of classical earth mechanics (except those indicated in Appendix 3).

The celestial geocentric reference frame (also called Earth-centred inertial) has for origin Earth's centre of inertia and as axes three axes directed toward fixed stars. This reference frame can be considered as inertial in a second approximation, better than the one of the reference frames attached to the Earth. The celestial geocentric reference frame is used to study phenomena affected by Earth's rotation and movements of Earth's artificial satellites.

Other inertial reference frames are presented in Appendix 3, with estimated orders of magnitudes of the committed approximations.

3 Gravitation and *Gravity*

Newton introduced in 1687 the theory of universal gravitation that specifies that two bodies attract each other in direct function of the product of their mass and in inverse function of the square of the distance separating their centres of mass. One recalls the following definitions:

> Gravitation is the physical phenomenon of attraction of a mass towards another mass.
> Gravity is the force associated with the phenomenon of gravitation.

(Note that here, the word 'Gravity' is not italicized.)

In a geocentric reference frame, the vectorial attraction force $\vec{F}_{gr}$ exerted by Earth of mass M_E on a proof mass m located at a vectorial distance $\vec{r}$ from the centre of the Earth reads:

$$\vec{F}_{gr} = -\frac{GM_E m}{r^2}\left(\frac{\vec{r}}{r}\right) \qquad F_{gr} = \frac{GM_E m}{r^2} \qquad (2)$$

where the negative sign indicates that the gravitational attraction force of the Earth on the mass m is directed in the opposite direction of the unit radius vector $\left(\frac{\vec{r}}{r}\right)$ and where F_{gr} is the norm of the vectorial force. G is a constant, called the gravitational constant, measured for the first time in 1798 by the English physicist Cavendish with laboratory experiments on attracting masses suspended to a torsion balance. The actual SI value of this constant is $G = 6.674\,28 \times 10^{-11}$ m^3 kg^{-1} s^{-2}.

Anticipating the conclusion, it is important to realise the following point: gravitation can never disappear! Gravitation and gravity are omnipresent in our universe and it is physically impossible to nullify one or the other. Never can you say that gravity is null! One must insist strongly on this important point that is unfortunately the source of much confusion: it is impossible to make gravitation disappear. The force of gravity never disappears! (except the two mathematical limit cases of a null mass, which is obvious, or an infinite distance between two bodies, which physically does not make any sense).

Before defining *gravity* and weight, let us take the following example. Let us consider an observer on the surface of the Earth which rotates on itself. This observer, wishing to determine the value of gravity, conducts an experiment to measure the force of attraction of Earth's mass on a proof mass. Despite all his efforts, he would only obtain a measure of the force resulting from gravity and the centrifugal force due to Earth's rotation, much smaller than gravity (about 300 times smaller at the equator), but still existing. This little centrifugal force is in fact an inertia force due to Earth's rotation, more precisely due to the fact that the experimental measurement is conducted in a non-inertial reference frame, the laboratory on Earth's surface (see Appendices 3 and 6). This global phenomenon that our observer on Earth can directly study is the phenomenon of *gravity*.

Gravity is the phenomenon resulting from the phenomenons of gravitation and inertia.
Weight is the force associated with the phenomenon of *gravity*.

Defined in a non-inertial reference frame, weight (the force due to *gravity*) is thus the sum of gravity (the force due to gravitation) and all inertia forces acting in this non-inertial reference frame (in the case of Earth, mainly those due to Earth's rotation). One will remember the following associations:

Reference frame:	inertial	non-inertial
Inertia forces:	absent	existing
Phenomenon:	gravitation	*gravity*
Associated force:	gravity	weight

Note that the word 'gravity' appears twice in this Table, once in italic and once in normal characters. It shows that this word designates two close, but different, concepts (see Appendix 1).

Before introducing the notion of weightlessness in the second chapter, we still have to discuss the Equivalence Principle, coming back first to the notion of mass of a body. Mass, defined above as the measure of the quantity of matter in a body, has two properties ensuing from the Principle of Inertia and from Newton's gravitation theory.

Firstly, from the Principle of Inertia it comes that the mass m of a body in relation (1) represents the inertia of this body, i.e. the measure of the resistance that this body opposes to its acceleration (i.e. to all modifications of its velocity); this mass is called inert or inertial mass and is noted further m_{in}.

Secondly, from Newton's gravitation theory, it comes that the mass m of a body in relation (2) represents the measure of attraction of this body to another one; this mass is called gravitational mass and noted further m_{gr}.

Newton admitted implicitly and without explaining it, that both inertial and gravitational masses were equal:

$$m_{in} = m_{gr} \tag{3}$$

This equality was experimentally confirmed up to a high degree of precision. At the beginning of the 20th century, Einstein made of this equality a founding base of general relativity. He saw in the equality of the inertial and gravitational masses a more profound reason, a fundamental equivalence between the effects of gravitation and the effects of inertia in an accelerated frame (i.e. a non-inertial frame). The equality of the inertial and gravitational masses must be interpreted as an essence identity between the forces of inertia and of gravity. This new concept is expressed in the

---

Principle of Equivalence

The phenomenons of gravitation and inertia are equivalent in a non-inertial reference frame.

Locally in a uniform gravitational field, a non-inertial reference frame can always be found such that the sum of inertia forces exactly balances the force of gravity.

This identity also means that:

Locally the two gravitational and inertial forces have indiscernible actions as they both yield to a mass an acceleration independent from this mass.

Equating relations (1) and (2) yields, with the equality (3):

$$F_{in} = F_{gr} \Rightarrow m_{in}\, a = \frac{G\, M\, m_{gr}}{r^2} \Rightarrow a = \frac{G\, M}{r^2} \tag{4}$$

The force of gravity of a mass M exerted on a body of mass m yield thus an acceleration a, independent from the mass m of this body.

In the particular case of a mass m on the surface of the Earth of mass M_E, one usually notes this acceleration a by the letter g. We will note it here g_{gr} (for gravitational g) to avoid confusing it with the acceleration of *gravity*, that will be noted g_w that is introduced in Appendix 6.

From (4), one obtains the expression of g_{gr}, the norm of the vectorial acceleration of gravity:

$$g_{gr} = \frac{G\,M_E}{r^2} \tag{5}$$

Let us insist again on the local character of the Principle of Equivalence, the identity between gravitational and inertial forces is only applicable locally, in a region where the value of the acceleration g_{gr}, function of the distance r, is not likely to change.

As already mentioned, all attempts to measure the acceleration g_{gr} of gravity conducted on the surface of Earth will always end up in the measurement of the acceleration g_w of *gravity*, i.e. the acceleration resulting from the acceleration g_{gr} of gravity and the acceleration caused by the inertial centrifugal force due to the Earth's rotation (and by other smaller inertial forces). The value of this acceleration g_w of *gravity* is on average 9.81 m/s^2 on the surface of the Earth. Appendix 6 summarize the distinctions to bring to the various values of g_w in different locations on Earth.

Conclusion of the First Chapter

Before answering to the question in the title of this chapter, it must be said that it is asked in a non-inertial reference frame. The answer is then affirmative, as inertia forces are one of the principal cause of motion in a non-inertial reference frame, consequence of inertia, the property of all masses to oppose changes in velocity.

The second chapter will answer to another question, namely is weightlessness without gravity?

Appendix 1: A Matter of Words

As seen in Sect. 3, the use of the single word 'gravity' can be confusing. So let us summarize what we mean exactly by gravity and *gravity*.

Gravity (without italics) is the force associated with the phenomenon of gravitation as explained in Sect. 3. *Gravity* (in italics) is the phenomenon resulting from the phenomenons of gravitation and inertia, and weight is the force associated with the phenomenon of *gravity*.

The common use of the single word 'gravity' to designate both concepts is the source of many confusions since centuries and its back translation in French and other languages has created additional confusion.

Another word exists in the English language, and although rarely used, it can exactly describe what we mean by *gravity* here. This word is weightiness. If we look at its definition, we find[2] "The state or quality of being physically heavy or weighty,

---

[2]Webster Dictionary, http://www.thefreedictionary.com/Weightiness,
  Princeton's WordNet, http://www.definitions.net/definition/weightiness,
  Webster 1913, https://www.niftyword.com/dictionary/weightiness/.

or the property of being comparatively great in weight; weight; force; importance; impressiveness".

So, we can extend the definition of *gravity* (in italics) given in the middle of Sect. 3 to the word weightiness, as the phenomenon resulting from the phenomenons of gravitation and inertia. Weight is the force associated with the phenomenon of weightiness.

However, the use of the word 'gravity' has become so widespread in the English language to describe the two concepts of gravitational force and the weightiness phenomenon that we will conform to this usage. Nevertheless, we will italicize 'gravity' to signify that we mean weightiness.

Let us note also that other languages have two different words for these two concepts, for example in French, '*pesanteur*' and '*gravité*' respectively for weightiness and gravity.

Appendix 2: Mass and Quantity of Matter

Newton defined mass as the measure of quantity of matter contained in a body and expressed as the product of the volume by the specific mass of the body. This definition should not be confused with another notion, called amount of substance that is defined as the number of elementary particles constituting a body (atoms, molecules, fractions of molecules, …) and whose unit is the mole. By convention, the amount of carbon ^{12}C whose mass is 0.012 kg represents an amount of substance of 1 mole, which is approximately equal to $6.02214199 \times 10^{23}$.

Furthermore, historically, what is called nowadays mass used to be designated formerly by '*quantitas materiae*' in Latin, which also contributed to the confusion.

Appendix 3: Inertial Reference Frames

To determine a perfect inertial reference frame is impossible. However, the principle of inertia is verified daily at our spatial and temporal scale. The concepts of inertial reference frame and of the inertia principle are tools of mathematical mechanics that allow to apprehend and to describe most physical phenomenons observed at macroscopic scale. This notion of model is important as it gives a description (most of time, a simple one) of an observed physical phenomenon with certain approximations. Practically, one tries to determine at best a reference frame supposedly inertial and adapted to the study of a specific problem with the best possible approximation, such that the error committed because of this approximation would be negligible with respect to the studied phenomenon, i.e. below a threshold arbitrarily fixed to a small fraction of the result obtained while studying this phenomenon.

The committed approximation is generally due to two inertial accelerations, expressed in a reference frame R in a non-uniform rectilinear motion with respect to another reference frame R_o, supposedly fixed (i.e. whose "inertial" character is better that the one of R). To evaluate an order of magnitude, let us suppose that the motion of R with respect to R_o is a pure rotation with a constant angular velocity ω. The vectors of centrifugal acceleration $\overrightarrow{a_e}$ and Coriolis acceleration $\overrightarrow{a_c}$ of a material point in R have norms

$$a_e = \omega^2 r \qquad a_c = 2\omega v_r \qquad \text{(A3.1)}$$

where r and v_r are the distance and relative speed of the point with respect to the origin of R (and assuming further that the displacement acceleration of the origin of R with respect to R_o is negligible and that the relative motion of the point is perpendicular to the rotation axis of R with respect to R_o; a course of Mechanics will provide more details, see books suggested in references).

Four reference frames are generally considered as shown in Fig. 2 and one proceeds by successive approximations.

1 Copernicus Reference Frame (R_C)

This reference frame has for origin the centre of inertia (or centre of mass) of the Solar System and for axes three orthogonal axes whose direction are fixed with respect to fixed stars. The reference frame can be considered in excellent approximation as an inertial reference frame. However, it is not exactly inertial as its origin is attached to the centre of inertia of the Solar System which is not fixed in the universe. Furthermore, the fixed stars toward which the three axes are directed are only fixed for a short instant at the astronomical scale, although sufficiently long to study the motion of planets since man observes them.

Instead of the Copernicus reference frame, one could consider a galactic reference frame (R_{Gal}) attached to the centre of the galaxy and whose axis' direction would be aligned on extra-galactic stars. Our galaxy describes a slow rotating motion such as the Sun, at 32 000 light-years ($\approx 3.03 \times 10^{20}$ m) from the galactic centre, describes a galactic revolution in 240 million years on average.

This galactic reference frame would be better (i.e. more "inertial") than Copernicus' one, but with the inconvenience that it would be extremely difficult to describe the motion of the Sun and planets, and, a fortiori, a phenomenon at the surface of the Earth. Furthermore, this galactic reference frame would not be perfectly inertial as well as our galaxy takes part in the general motion of the local cluster of galaxies and of universe expansion.

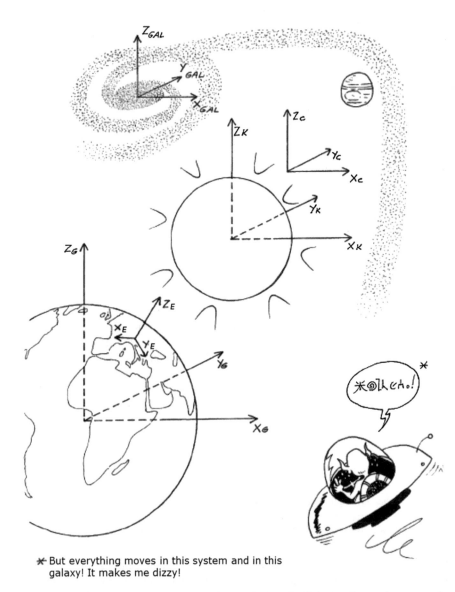

✳ But everything moves in this system and in this
galaxy! It makes me dizzy!

Fig. 2 Artist's impression of some inertial reference frames: the Galactic reference frame R_{Gal} (in galaxy's centre), Copernicus reference frame R_C (between the Sun and Jupiter), Kepler reference frame R_K (in the Sun's centre), the celestial geocentric reference frame R_G (in Earth's centre) and a reference frame attached to Earth's surface R_E. (Credit: P. E. Paulis)

Calculating by (A3.1) the committed approximation by considering Copernicus reference frame instead of the galactic reference frame, the orders of magnitude

Table 1 Approximations of inertial reference frames

	r (m)	ω (rad/s)	a_e (m/s^2)	a_c (m/s^2)
R_C/R_{Gal}	r_{Gal} $= 3.03 \times 10^{20}$	ω_{Gal} $= 8.3 \times 10^{-16}$	2.09×10^{-10}	1.7×10^{-15}
R_K/R_C	$r_{C.i.S\text{-}J}$ $= 7.43 \times 10^8$	$\omega_{J.rev.}$ $= 1.678 \times 10^{-8}$	2.1×10^{-7}	3.4×10^{-8}
R_G/R_K	$r_{E.orb.}$ $= 1.496 \times 10^{11}$	$\omega_{E.rev.}$ $= 1.991 \times 10^{-7}$	5.93×10^{-3}	4×10^{-7}
	$r_{C.i.E\text{-}M}$ $= 4.67 \times 10^6$	$\omega_{M.rev.}$ $= 2.662 \times 10^{-6}$	3.3×10^{-5}	5.3×10^{-6}
R_E/R_G	$r_{E.eq.}$ $= 6.37814 \times 10^6$	$\omega_{E.rot.}$ $= 7.292 \times 10^{-5}$	3.39×10^{-2}	1.5×10^{-4}

Inertial centrifugal (a_e) and Coriolis (a_c) accelerations (for which $v_r = 1$ m/s was assumed).

shown in Table 1 are perfectly negligible in front of the value of $g_w = 9.81$ ms^{-2} (one considered a unit velocity of 1 m/s for the relative speed v_r in Coriolis' acceleration).

2 Kepler Reference Frame (R$_K$)

This reference frame has for origin the centre of inertia of the Sun and axes parallel to those of the Copernicus reference frame. This heliocentric reference frame can be considered with a very good approximation as an inertial reference frame, as the Sun centre of inertia is always close to the Solar System centre of inertia.

As the ratio of all planet masses to the Sun's mass is about 1.3×10^{-3} and that the ratio of Jupiter to Sun's masses is about 9.55×10^{-4}, one can approximate the distance between the origins of Kepler and Copernicus reference frames by the distance between the centre of inertia of the reduced Sun-Jupiter system (C.i.S-J) and the Sun's centre. The C.i.S-J is at about 7.43×10^8 m from the Sun's centre on the axis Sun-Jupiter, just outside the Sun whose radius is 6.96×10^8 m. Considering that the C.i.S-J moves with the orbital velocity of Jupiter (whose orbital eccentricity is neglected), one can evaluate the committed approximation in the Kepler reference frame with respect to Copernicus reference frame. The orders of magnitude shown in Table 1 are also negligible. The largest distance from the Solar System centre of inertia in the most unfavourable theoretical case in which all the planets are in conjunction on the same side of the Sun and aligned, i.e. passing through their orbit node. In this configuration, the Solar System instantaneous centre of inertia is at 1.51×10^9 m from the centre of the Sun and the value of a_e in Table 1 is only doubled. Any other planetary configuration would yield a smaller value.

3 Earth's Reference Frames

In Kepler's reference frame, Earth's trajectory around the Sun is modelled by an elliptical orbit having the Sun in one of its foci. In reality, Earth's motion is more complicated and can be decomposed as follows:

(1) The centre of inertia of the Earth-Moon system describes a quasi-elliptical trajectory around the centre of inertia of the Solar System;
(2) The Earth's centre of inertia describes a quasi-elliptic trajectory around the centre of inertia of the Earth-Moon system;
(3) Earth describes a complex rotation motion around its centre of inertia, that can be represented by a rotation movement around an axis, affected itself by several very slow movements, the mains being precession and nutation.

Therefore, Earth's reference frames (the celestial geocentric reference frame or a reference frame attached to Earth) are far from being inertial!

3.1 Celestial Geocentric Reference Frame (R_G)

Nevertheless, the celestial geocentric reference frame (with an origin at Earth's centre of inertia and axes parallel to those of Kepler's frame) can be considered as inertial with a good approximation, due to the motions given in (1) and (2) above.

This approximation is generally sufficient to study Earth's artificial satellites' motion, Eastward deflection of falling bodies, gyroscopes, Foucault's pendulums, winds and ocean currents, but not tides.

Considering that motion (1) can be approximated by Earth's sidereal revolution with angular velocity $\omega_{E.rev.}$ on a circular orbit of radius $r_{E.orb.}$ (eccentricity $e = 0.017$ is sufficiently small to be neglected), one finds in Table 1 the orders of magnitude of inertial centrifugal and Coriolis accelerations due to motion (1) above.

Motion (2) can be analysed as in the case of the reduced Sun-Jupiter system here above. Knowing the Earth-Moon distance (3.844×10^8 m) and the ratio of their masses (≈ 81.3), the centre of inertia of the Earth-Moon system (C.i.E-M) is at 4.67×10^6 m from the Earth's centre, or at approximately 3/4 of Earth's radius in direction of the Moon. Assuming a circular Moon's orbit (its eccentricity $e = 0.0549$ can be neglected), one can say that Earth's centre of inertia describes a rotation around the C.i.E-M at the Moon sidereal orbital angular velocity ($\omega_{M.rev}$). Can we neglect the effect of this offset rotation of the Earth around the inertia centre of the Earth-Mon system? The answer, surprisingly in view of the distance (3/4 of Earth's radius) is yes: the inertial accelerations given in Table 1 are negligible.

A similar reference frame, called Earth Centred Inertial (ECI) reference frame, is used to describe orbits and trajectories of artificial satellites and spacecraft in astronautics. Its origin is at the Earth's centre of inertia. The fundamental plane, formed by the X and Y axes, includes Earth's equator and the X axis points toward the vernal equinox (see Fig. 3).

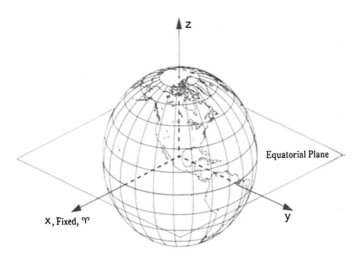

Fig. 3 The Earth-Centred Inertial (ECI) reference frame (from Ref. 7)

3.2 Reference Frames Attached to Earth (R_E)

The reference frames attached to Earth (with an origin at the centre or at the surface of the Earth and axes fixed with respect to Earth, assumed solid and not deformable) can also be considered as inertial with an approximation not as good as above, but still sufficient to study local phenomenons on Earth. In this approximation, one neglects the inertial centrifugal and Coriolis accelerations due to the rotation motion (3) above. Considering only Earth's sidereal rotation around its axis at the angular speed $\omega_{E.rot}$, the inertial centrifugal acceleration at the equator is approximately 300 times less than the acceleration g_{gr} of gravity. One can thus generally neglect it in a first approach; however, it should be taken into account for precise measurements (see Appendix 5).

One of the reference frames attached to Earth that is used commonly is the Earth Centred, Earth Fixed (ECEF) reference frame, with the X axis always aligned with the meridian of Greenwich which is defined as longitude 0° (see Fig. 4).

Appendix 4: Fundamental Forces and Inertial Forces

Except for inertia forces, all forces in the universe are based on four fundamental interactions, namely gravitation, weak, electromagnetic and strong interactions.

The gravitational force acts between masses and is an attractive force. The electromagnetic force acts between electric charges, and can be attractive between charges of opposite signs or repulsive between charges of same signs. The strong and weak forces are nuclear forces responsible for the interactions between subatomic particles

Fig. 4   The Earth Centred,
Earth Fixed (ECEF)
reference frame (from Ref.
29)

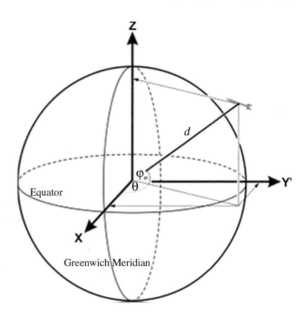

Fig. 4   The Earth Centred, Earth Fixed (ECEF) reference frame (from Ref. 29)

and acting only at very short distances. The strong nuclear force is the force responsible for the structural integrity of atomic nuclei, allowing the binding of the protons and neutrons of the nuclei, while the weak nuclear force is responsible for the decay of certain nucleons into other types of particles. The most known effect is the beta decay of neutrons inside the nucleus of atoms and the radioactivity produced by this decay. The standard model of particle physics predicted the unification of the weak and electromagnetic forces in the electroweak theory which observation has confirmed. The unification of the other forces is a current active field of research in physics.

All other forces in nature derive from these four fundamental interactions. For example, friction forces and the spring recall force are manifestations of the electromagnetic force.

On the other hand, inertia forces appear only in non-inertial reference frames. These include centrifugal forces and Coriolis forces and more generally, any forces that appear in a non-inertial reference frame in motion with a non-uniform velocity with respect to an inertial reference frame. These inertia forces are not genuine, like the four fundamental forces above, as they depend on the reference frames.

Inertial forces are sometime called "pseudo-" forces or "fictitious" forces by some authors. According the Cambridge online dictionary[3], "pseudo-" means "pretended and not real" and "fictitious" means "invented and not true or existing; false". This choice of words, "pseudo-" or "fictitious", is very unfortunate, as inertial forces are "real", although it is clear that they are not similar to the four fundamental forces deriving from fundamental interactions.

---

[3]https://dictionary.cambridge.org/dictionary/english/.

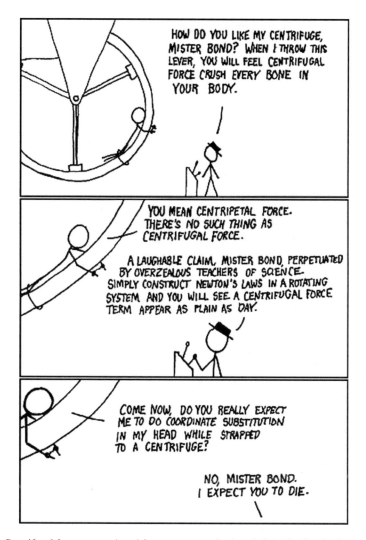

Fig. 5 Centrifugal force vs centripetal force: a matter of point of view (Credit: xkcd)

Let us take a simple example. If the passenger walking in the train of Fig. 1 falls due to the acceleration or deceleration of the train (cases 2 and 3), the lump of his forehead will be real and painful, and not "fictitious" or "pseudo-" …

Therefore, to differentiate fundamental forces from forces appearing in non-inertial reference frames, one could call these changing-reference-frame forces, or accelerated-reference-frame forces, or simply… inertial forces (Fig. 5).[4]

---

[4]https://xkcd.com/123/.

Appendix 5: How to Measure and Express Weight?

Defined in a non-inertial reference frame, the weight W of a body is a vectorial entity, defined as the product of the body mass by the vectorial acceleration of *gravity* g_w

$$\vec{W} = m\vec{g}_w \qquad (A5.1)$$

The acceleration of *gravity* g_w is further defined in Appendix 6.

One measures usually the weight of a body in two ways, with a scale or with a spring apparatus. At the risk of tiring the reader, let us recall how to use a scale. The body whose weight we want to measure is placed on one of the scale platforms. One places on the other platform one or several mass standard(s), which are calibrated and marked masses (and that the common language designates unfortunately and incorrectly by the word "weight"). When the balance beam comes to horizontal by adding or removing a certain number of mass standards, the sum of weights of these mass standards indicates the result of the weighting of the body.

When this method is applied in a uniform gravitational field (i.e. with a scale having sufficiently small dimensions compared to the radius of curvature of the Earth gravitational field), this method comes down to compare two masses:

$$W_{body} = W_{stand.} \Rightarrow m_{body}g_w = m_{stand.}g_w \Rightarrow m_{body} = m_{stand.} \qquad (A5.2)$$

Conducted on the surface of another planet, in a different gravitational field, this method always gives the same result, as the local acceleration of *gravity* g_w applied two both platforms of the scale is equal (the simplification by g_w on both side of the second equality of (A5.2) is independent from the value of g_w). Let us note that this method, used since antiquity, has contributed to the confusion between notions of weight and mass, the equality of ones yielded from the equality of the others, but not implying the equality between them. The use of the term 'weight' to designate mass is a misnomer and is unfortunately still very common.

The other method is a dynamic method, using apparatus like extension or compression dynamometers, the most common being the bathroom scale. This method does not compare two masses but two forces, the force associated to *gravity* (the weight) and the force due to the extension of compression of a graduated spring:

$$\vec{F}_{spring} = -k\vec{x} \qquad (A5.3)$$

where k is a spring characteristic, called the spring constant or spring stiffness, $\vec{x}$ is the vectorial change of the spring length and the negative sign expresses that the spring force is applied in the opposite sense of the change of the spring length.

Conducted on another planet, in a different gravitational field, this method never gives the same result, as the weight depends on the local *gravity* while the spring force is independent from it. For example, using the same extension dynamometer on Earth and on the Moon, one would have at equilibrium:

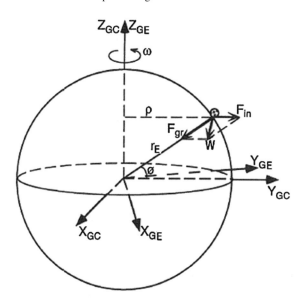

Fig. 6 One reasons in the non-inertial reference frame attached to the Earth R_{GE} or in the laboratory reference frame (not represented), in rotation with respect to the celestial geocentric reference frame R_{GC}, considered as inertial. The body to be weighted is on Earth's surface at a point of latitude ϕ and at a distance ρ ($= r_E \cos\phi$) from the rotation axis. The body weight W is the vectorial sum of F_{gr} and F_{in} (whose magnitude has been exaggerated on the drawing for visual clarity). The deviation angle φ between the directions of W (direction of the local vertical) and of F_{gr} (direction of Earth radius) is also exaggerated on the drawing

$$W = F_{spring} \Rightarrow \begin{cases} mg_{E.w} = kx_1 & \text{(on Earth)} \\ mg_{M.w} = kx_2 & \text{(on the Moon)} \end{cases} \qquad (A5.4)$$

with $x_1 > x_2$, as $g_{E.w} > g_{M.w}$, Moon's *gravity* being approximately six times less than Earth's *gravity*.

Let us conduct now the body weighting experiment with a dynamometer or with a bathroom scale on the surface of the Earth. Which reference frames are we going to consider? In the celestial geocentric reference frame (assumed inertial), the reference frame of the laboratory is non-inertial as it is dragged along by Earth's rotation (see Fig. 6). This rotation induces in the laboratory reference frame an inertial centrifugal force whose norm is

$$F_{in} = F_{centrif.} = ma_{centrif.} = m\omega_{E.rot.}^2 \rho = m\omega_{E.rot.}^2 r_E \cos\phi \qquad (A5.5)$$

where ρ is the distance from the point where the weighting is done to the Earth's rotation axis. Assuming a spherical Earth of radius r_E and the laboratory located at a place of latitude ϕ, one obtains the last equality in (A5.5).

This force is directed perpendicularly to the Earth's rotation axis and is maximal on the equator ($\phi = 0°$) and nil at the poles ($\phi = 90°$). In the laboratory's non-inertial reference frame (or any reference frame attached to Earth, whether geocentric or not), the weight is the vectorial sum of the force of gravity and of the inertial centrifugal force. The weight vector is not directed toward the centre of the Earth (assumed spherical) as the gravitational force, but along the vector resulting from the vectorial addition of the force of gravity and the inertial centrifugal force (see Fig. 6). This direction is the local vertical, given by the plumb line, and deviating by an angle φ from the direction of the local earth radius. The weight vector and its norm read

$$\vec{W} = \vec{F}_{gr} + \vec{F}_{in} \Rightarrow W = F_{gr} \cos \varphi - F_{in} \cos (\phi + \varphi) \approx F_{gr} - F_{in} \cos \phi \quad \text{(A5.6)}$$

where the gravity force and the inertial centrifugal force are projected on the direction of $\vec{W}$. The deviation angle φ is nil (i.e. the local vertical is along the earth radius) at the equator (where the gravity and inertial centrifugal forces are aligned but in opposite senses) and at the poles (where the inertial centrifugal force is nil). For other latitudes, one shows that this angle φ is small, the maximum value is less than $0.2°$ arising at latitudes $\phi = \pm 45°$ (for a spherical model of the Earth) and decreases when coming closer to the equator or to the poles. One can then neglect φ, like in (A5.6). Note as well that the negative sign in front of F_{in} shows that this inertial force must be subtracted from the gravity force. Replacing the norms of W, F_{gr} and F_{in} by (A5.1) and (A5.5) yield

$$W = mg_w = m \left(\frac{GM_E}{r_E^2} - \omega_{E.rot.}^2 r_E \cos^2 \phi \right) \quad \text{(A5.7)}$$

This expression of the weight is correct in its principle, but it can still be corrected for the values of r_E and $\omega_{E.rot.}$ (see Appendix 6). The inertial centrifugal force due to Earth's rotation is implicitly measured in the experiment of measuring the weight conducted in a non-inertial reference frame attached to the Earth, geocentric or not.

One will recall that the mass of a body is invariable and stay the same anywhere in the universe (if there are no nuclear reactions in the body and if one can neglect relativistic effects), while the weight depends on the location in the universe.

Appendix 6: The Acceleration g

The acceleration g_{gr} defined in relation (5) is indeed the acceleration of gravity. However, by an unfortunate but current language abuse, the average value of 9.81 ms^{-2} given at the end of Sect. 3 is in reality the average value of the acceleration of *gravity*, expressed in a reference frame attached to the Earth and considered as non-inertial. As shown in Appendix 5 in (A5.7), the average acceleration of *gravity* in a point on Earth's surface of latitude ϕ can be written in a first approximation as

Table 2 Approximate values of gravity accelerations

	$g_{gr} = (ms^{-2})$	$a_{centrif.} = (ms^{-2})$	$g_w = (ms^{-2})$
At equator ($\phi = 0°$)	9.80	0.04	9.76
At the poles ($\phi = 90°$)	9.86	0	9.86
On average ($\phi = \pm45°$)	9.83	0.02	9.81

$$g_w = g_{gr} - a_{centrif.} \cos\phi = \frac{GM_E}{r_E^2} - \omega_{E.rot.}^2 r_E \cos^2\phi \qquad (A6.1)$$

The accelerations g_w and g_{gr} given here above depend on Earth's radius, i.e. on the distance of Earth's attractive centre to a point located on Earth surface. This distance varies in function of latitude ϕ as Earth is not exactly spherical but can be represented in a second approximation by an ellipsoid of revolution flattened at the poles. The values of Earth's equatorial and polar radii recommended by the International Astronomical Union are respectively 6 378 136.6 and 6 356 752.3 m. Relation (A6.1) yields then the values in Table 2.

These values are approximate but generally sufficient for calculations in the first order.

Chapter 2
Is Weightlessness Without Gravity?

Introduction to the Second Chapter

In Chapter 1, we have introduced the notions of inertia forces that exist in non-inertial reference frames in accelerated motion with respect to an inertial reference frame, in which there are no inertia forces. *Gravity* (weightiness) has been defined as the phenomenon resulting of gravitation and inertia. In a non-inertial reference frame, the force associated to *gravity* (weightiness), weight, is the sum of the gravity force and of all inertia forces.

In this chapter, the notions of free fall and weightlessness are defined in Sect. 4. Free fall trajectories are presented in Sect. 5 and free fall in the universe is addressed in Sect. 6.

4 Free Fall and Weightlessness

What gives weight to a body are gravity and the centrifugal inertia force due to Earth's rotation, as seen previously. On the other hand, the sensation of weight, what the sole of the foot feels or what the bathroom scale indicates, is only possible thanks to the obstacle on which rest the feet or the bathroom scale: the ground or the bathroom floor. Imagine for a moment that the ground suddenly disappears or that we fall through a hole in the floor, one loses immediately the sensation of weight because there is nothing to cause it anymore: one falls freely.

Let us continue to imagine that the fall is performed now in a well or in a tower. Air friction is such that the fall velocity would rapidly reach a constant value. Let us imagine further that vacuum has been made in this well or tower, i.e. all air has been pumped out. There is nothing anymore that would oppose a perfect free fall. There is no force anymore acting on the body in free fall, except for gravity of course, as the latter is responsible for the movement of free fall.

If one accepts that a reference frame attached to Earth's surface would be inertial, the movement of free fall with respect to this reference frame is rectilinear and uniformly accelerated. On the other hand, in the reference frame attached to the

© The Author(s) 2018
V. Pletser, *Gravity, Weight and Their Absence*, SpringerBriefs in Physics,
https://doi.org/10.1007/978-981-10-8696-0_2

body that falls freely, the sensation of weight has disappeared, *gravity* (weightiness) is null, the body does not weight anymore although obviously, it keeps the whole of its mass.

An important consequence of the Equivalence Principle (see Chapter 1) allows us to introduce at last the notion of weightlessness. As there is equivalence between the phenomena of gravitation and inertia in a non-inertial reference frame attached to a body in a perfect free fall, one can always find such an accelerated non-inertial reference frame that allows us to modify or nullify locally and temporarily the effects of a gravitational field, i.e. a non-inertial reference frame in which the sum of inertia forces equals gravity in magnitude but in opposite direction. "Locally" means here only in this region of the universe corresponding to the vehicle or the body in free fall, and "temporarily", only for the duration of the free fall.

Let us take the example of an observer in a lift cabin without any windows. He has no outside visual clues. The observer holds an object in hand that he releases without throwing it (i.e. without initial velocity). Although air friction in the lift shaft is always present, one accepts that it is negligible during the cabin motion (its velocity is sufficiently small allowing to neglect this small perturbing force).

Consider the case in which the lift cabin is going up with a certain acceleration. The reference frame attached to the lift is thus non-inertial: with respect to this reference frame, the observer's weight increases slightly; it equals his normal weight on Earth (directed toward the floor), plus the inertia force due to the accelerated movement of the lift cabin with respect to a terrestrial reference frame; this inertia force is directed downward, in the opposite direction of the upward acceleration of the cabin.

Consider now the following three cases in which the lift cabin is going down as shown in Fig. 7.

(1) The cabin goes down with an acceleration a_L smaller than g_w: the elevator functions normally. The reference frame attached to the cabin is not inertial: with respect to this reference frame, the observer's weight decreases slightly; it equals his normal weight on Earth (directed toward the floor), plus the inertia force due to the accelerated movement of the lift cabin with respect to a terrestrial reference frame; this inertia force is directed upward, in the opposite direction of the downward acceleration a_L of the cabin (see the minus sign in relation (1) in Chapter 1). If the observer releases the object that he held, it falls toward the floor.

(2) The cabin is "pushed" downward and descends with an acceleration a_L larger than the acceleration g_w of *gravity* (weightiness). The observer suddenly has a "negative weight" with respect to the reference frame attached to the cabin, i.e. he rises to the ceiling and bumps his head with a force that is the resultant between his normal weight on Earth (directed toward the floor) and the inertia force due to the downward accelerated movement of the lift cabin with respect to a terrestrial reference frame; this inertia force is directed upward, in the opposite direction of the downward acceleration a_L of the cabin. If the observer lets the object go from his hand, it also rises to the ceiling.

Fig. 7 Example of an observer in a lift cabin descending with an acceleration a_L; the cabin reference frame R_L is non-inertial; the reference frame R_E on ground is considered as inertial. Three cases: (1) normal descent: $a_L < g_w$ (left); (2) forced descent: $a_L > g_w$ (middle); (3) descent in free fall: $a_L = g_w$ (right). (Credit: P.E. Paulis)

(3) The cabin goes down with an acceleration a_L exactly equals to the acceleration g_w of *gravity* (weightiness). It is the typical case of rupture of the retaining cable of the lift cabin: the cabin falls freely. The observer and the released object float in the cabin, without going up or down in the cabin. The observer has a "null weight" with respect to the cabin reference frame as it is the difference of his normal weight on Earth (directed toward the floor) and the inertia force of the cabin motion with respect to a terrestrial reference frame; this inertia force is directed upward, in the opposite direction of the cabin free fall. These two forces are equal and in opposite directions, they cancel each other and the observer does not weight anything anymore. He is falling with the same velocity and the same acceleration as the cabin and all its content. The object released from his hand floats freely next to the observer in the cabin reference frame. Both are in a state of weightlessness in the reference frame of the cabin. The observer does not feel anymore the sensation of weight.

Consider now the point of view of an external observer standing on the ground. In his referential frame, that is attached to Earth's surface and that we suppose inertial, he sees the lift cabin falling in a uniformly accelerated movement (if one can also neglect air friction). There is thus no force other than gravity that acts and the external observer concludes that, with respect to his inertial reference frame, the cabin and its content are in free fall.

On the other hand, for the courageous observer in the cabin, the situation is slightly different. What can he logically deduce from his experience in the third case (other than it is time to call the repair service)? The observer, as a good scientist, proceeds by deduction of what he observes: he floats freely in the cabin reference frame. Two logical interpretations are available to him and they are important to understand:

(1)   either he deduces that his weight is null with respect to the cabin reference frame, that *gravity* (weightiness) is cancelled in his cabin and that, in the reference frame attached to the cabin, he is in a state of weightlessness. Consequently, he deduces that the cabin is in free fall in a gravitational field and that there is no other force than gravity applied to him;

(2)   or, alternatively, in absence of any other information from the outside world (there are no windows in the cabin), he can deduce that, gravity seems to no longer exist in his cabin, as if Earth would not gravitationally attract it any longer. Suddenly, the attraction of the mass of Earth and all the masses in the universe would have vanished. The reference frame of the cabin, from non-inertial, seems to have become inertial and the cabin movement seems to have become a uniform rectilinear translation movement. He does not feel any inertia force anymore.

Both interpretations can very logically be deduced from the observation made in the cabin without any other information. However, we know that gravitational attraction cannot be nullified; it would mean that all masses in the universe must be nullified or that one should be at an infinite distance of any mass, both being physically impossible. Therefore, the second interpretation must be rejected. These two interpretations are not simply a theoretical artefact or a pure construct of the mind. It is really a fundamental indiscernibility: a physics or chemistry experiment performed in a laboratory in a perfect free fall gives results identical to those that would obtained by nulling gravitational attraction, i.e. by doing "$g = 0$" in the equations of a theoretical model of a phenomenon. There is no physical difference between the consequences of a movement of free fall and the total absence of gravitational field.

One will remember that:

A body is in free fall if it is subjected to the only force of gravity in an inertial reference frame and this body and all its content are in a state of weightlessness in the non-inertial reference frame attached to the body.

Weightlessness thus appears in a non-inertial reference frame, which is in a state of free fall with respect to an inertial reference frame. Weightlessness is a dynamical state that requires a free fall movement.

It is also important to specify the reference frame in which one reasons. So, the observer in the non-inertial reference frame of the cabin in free fall with respect to an external inertial reference frame, is in the state of weightlessness resulting from the equality of the gravity force and of the inertia forces in this non-inertial reference frame (the cabin observer "sees" only the weightlessness). On the other hand, for an external observer in an inertial reference frame (e.g. an observer on the ground), the cabin is in free fall and he does not "see" the weightlessness (there are no inertia forces in his inertial reference frame): for him, the cabin and all its content are falling with the same acceleration (the external observer "sees" only the free fall). These two interpretations are not contradictory but complementary. They express a same phenomenon but from different points of view, in different reference frames depending on the situation of the observer.

One appreciates the paradox: a vehicle can be submitted to a free fall movement in a gravitational field, giving the passengers the impression that this gravitational field does not exist. To overcome gravity, do not resist it!

5 Free Fall Trajectories

So far, we have only considered the free fall motion along the descending vertical. How to explain weightlessness on board space vehicles in orbit around the Earth?

The notion of free fall can be generalized as follows:

> A body which is only subjected to the action of a gravitational field is in free fall.

The resultant of all forces other than gravity acting on the body (lift cabin, aircraft, spacecraft, satellite, ...) must be null, i.e. there is no aerodynamic drag, nor engine thrust, nor any disturbing forces.

In this generalization, the notion of free fall should not be understood as the simple downward undisturbed fall movement, but as any movement which is only due to the action of a gravitational field.

Generally speaking, the trajectory of a body moving in a uniform gravitational field and without any other force (no air drag, no engine thrust, etc.) is called ballistic and its shape is a conic (ellipse, circle, parabola or hyperbola) or a degenerated conic (straight line), solution of movement second degree equations. The type of trajectory depends only on the initial velocity (expressed in an inertial geocentric reference frame), at the moment when the action of all forces other than gravity stops, in particular propulsion at the time of the initial impulsion. Figure 8 shows some free fall trajectories.

If the initial velocity is null or directed downward, the free fall trajectory is the descending vertical.

If the initial velocity is directed upward (and smaller than escape velocity, approximately 11.2 km/s on Earth, see Appendix 7), the free fall trajectory is the ascending

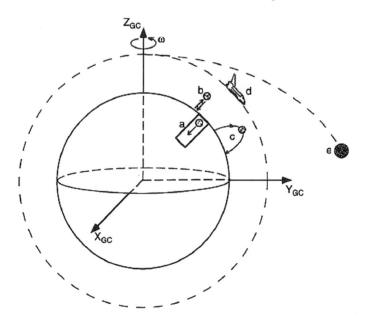

Fig. 8 Some free fall trajectories, depending on initial velocity v_0: **a** descending vertical for $v_0 = 0$; **b** ascending vertical for v_0 vertical upward and $0 < v_0 < 11.2$ km/s; **c** parabolic for v_0 inclined on the horizontal and $0 < v_0 < 11.2$ km/s; **d** circular orbit for $v_0 \approx 8$ km/s and horizontal v_0 (perpendicular to the straight line between the inertia centres of Earth and the object in orbit); **e** hyperbolic orbit for $v_0 > 11.2$ km/s

vertical until the moment when all the kinetic energy has been transformed into potential energy, after which the trajectory becomes again the descending vertical.

For a non-null initial velocity and smaller than escape velocity and whose direction is different from vertical, the body describes a ballistic trajectory before falling back on ground. This ballistic trajectory is a small arc of a very flat ellipse having Earth's centre at one of its foci. In first approximation (where one considers a parallel gravitational field instead of radial, i.e. in a region of space sufficiently small for the approximation to be acceptable), one can show that the ballistic trajectory is a parabola (see Appendix 7).

If air drag can be neglected (i.e. if we are sufficiently high) and if one gives a larger initial velocity to a body along the local horizontal (i.e. along a direction perpendicular to the radius between the inertia centres of Earth and of the body), the free fall trajectory is circular or elliptic, typical of satellites orbiting Earth. A satellite in orbit is also in free fall around the Earth, but because its sufficiently large velocity (in the order of 28 000 km/h, or approximately 8 km/s) along the perpendicular to the local Earth's radius, the satellite does not fall back directly on Earth but continues to move around the Earth. One needs of course to go up to several hundreds of kilometres above Earth's surface for the atmospheric drag to be really negligible.

Let us increase the initial impulsion in giving a larger initial velocity and there goes our body in free fall on a parabolic or hyperbolic orbit that would move it away from Earth for ever.

One can thus generalize our statement and say that:

For a body moving in a gravitational field, if the resultant of all non-gravitational forces applied to the body is null, this body and all its content are in free fall in an inertial reference frame and in a state of weightlessness in a non-inertial reference frame attached to the body.

Appendix 8 develops in more details which part of the body exactly is in weightlessness.

6 Free Fall in the Universe

So far, we have only considered Earth's gravitational field. If one considers gravitational fields of other planets and celestial bodies, trajectories can be modified, but the state of free fall remains as long as the body is only submitted to the action of the superposition of all gravitational fields.

This yields also that all spacecraft are in free fall state on their trajectory as long as their propulsion or attitude engines are not functioning and that there are no other perturbing forces. Planets, asteroids, and all celestial bodies whose movement is only caused by gravitation are also in free fall as they are only submitted to the gravitational attraction of the Sun and other planets and bodies.

Let us take the example of a spacecraft like Voyager on its way to planet Jupiter. The spacecraft is not continuously propelled, and as long as there are no other forces than gravity, Voyager is in free fall in space and all its components are in weightlessness inside the spacecraft.

On the way, its trajectory is modified and the spacecraft describes an abrupt turn, as illustrated in Fig. 9.

Let us consider two possible causes:

(1) In the first case, NASA technicians have ignited remotely a propulsion engine. Weightlessness stops immediately in the spacecraft as a non-gravitational force appears that induces an acceleration to the probe.
(2) In the second case, Voyager come close to an asteroid and its trajectory is modified in the same way. However, as there are no non-gravitational forces, only the action of the asteroid gravitational field the craft remains in free fall and the state of weightlessness persists inside the probe.

Surprising at first sight: effects apparently similar but caused differently yield radically opposite consequences. In the first case, a passenger in an interplanetary

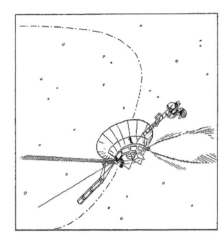 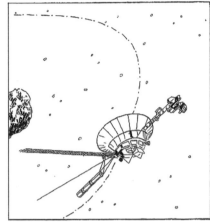

Fig. 9 The trajectory of an interplanetary probe curves because of (1) an engine switching on, then weightlessness ceases in the probe (left), or (2) proximity of an asteroid, then weightlessness persists in the probe (right). (Credit: P. E. Paulis)

vehicle would be squashed on its seat; in the second case, the passenger would continue to float freely.

A last example: near a black hole, the gravitational attraction is immense, quasi infinite. A neutral particle attracted gravitationally by the black hole and that is subjected to no other force is also in free fall. This example is of course very approximate as, according the back hole theory, no particle can stay neutral, and forces other than gravitational and relativistic effects cannot be neglected.

One understands that in the universe, free fall is the general rule and *gravity* (weightiness) the exception as it is encountered only at the surface of a celestial body or during an interaction with a force other than gravity. Let us generalize once more our statement

> If a body is moving in a gravitational field resulting from the superposition of attractions of other celestial bodies and the resultant of external non-gravitational forces applied to the body is null, this body and all its content are in free fall in an inertial reference frame and in a state of weightlessness in a non-inertial reference frame attached to the body.

Unfortunately, this simple notion is not often understood, even sometimes by eminent scientists. In Appendix 9, we refute some incorrect ideas that unfortunately are persistent.

Conclusion of the Second Chapter

We can now answer the question asked in the title of this second chapter. No, weightlessness is not without gravity as it is the cause of the free fall, *sine qua non* condition of weightlessness. In the third chapter, we answer to another question: should one weigh on microgravity?

Appendix 7: Ballistic Movement and Escape Velocity

A body thrown upwards with an initial velocity inclined on the horizontal describes a certain trajectory and falls back on Earth. One shows simply that this trajectory is parabolic if:

- the travelled horizontal distance is sufficiently small in front of the curvature radius of Earth's surface for the gravitational field (more precisely the *gravity* (weightiness) field) to be approximated by a parallel field, i.e. the direction of the vectorial acceleration g stays parallel to itself in all points of the trajectory; and
- the travelled vertical distance is sufficiently small in front of Earth's radius to consider that the gravitational field is uniform, i.e. that the magnitude of the acceleration g varies in a negligible way.

After the initial impulsion, if air drag is neglected, the only force that acts on the body is its own weight coming from the phenomenon of *gravity* (weightiness).

As the ballistic movement is described in a vertical plane that includes the whole of the trajectory, one chooses a local reference frame such that its origin is at the injection point (the point where the initial impulsion stops), the X axis horizontal, the Z axis along the ascending local vertical and the Y axis perpendicular to the vertical plane formed by the X and Z axes.

Projecting on the horizontal X and vertical Z axes the constant initial velocity v_0 making an angle θ with horizontal (see Fig. 10), one obtains a constant horizontal velocity v_x and a vertical velocity v_z function of time t:

$$v_x = v_0.\cos\theta = \text{constant} \tag{A7.1}$$

$$v_z = v_0.\sin\theta - g.t \tag{A7.2}$$

Integrating velocities with respect to time, one obtains the components along the X and Z axes of the body position:

$$x\,(t) = v_0.\cos\theta.t \tag{A7.3}$$

$$z\,(t) = v_0.\sin\theta.t - \frac{g.t^2}{2} \tag{A7.4}$$

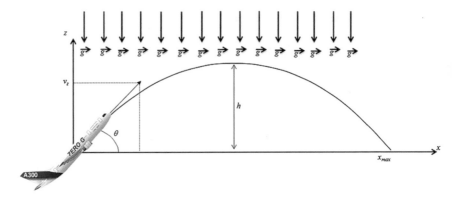

Fig. 10 Ballistic parabolic trajectory in a gravitational field considered as uniform

(where integration constants are null because of the choice of the local reference frame). Eliminating time between (A7.3) and (A7.4) yields the equation of the parabola shaped trajectory:

$$z\,(x) = -\frac{g}{2v_0^2.\cos^2\theta}.x^2 + x.tg\theta \qquad (A7.5)$$

The total duration T of the movement on the parabolic trajectory corresponds to the non-zero value of t in the second-degree Eq. (A7.4) in t^2 after posing $z = 0$ (i.e. when the parabola intersects again the X axis):

$$T = \frac{2v_0.\sin\theta}{g} \qquad (A7.6)$$

As the parabolic trajectory is symmetrical, the trajectory maximum height h is reached for $t = \frac{T}{2}$ in Eq. (A7.4), yielding

$$h = \frac{v_0^2.\sin^2\theta}{2g} \qquad (A7.7)$$

The distance d travelled horizontally during the parabola is found from Eq. (A7.3) by replacing t by T, yielding

$$d = \frac{v_0^2.\sin 2\theta}{g} \qquad (A7.8)$$

and is maximum for $\theta = 45°$.

These equations define the ballistic movement, i.e. unpropelled. Let us note that the duration, the horizontal distance and the maximum height of this ballistic movement are independent from the body mass.

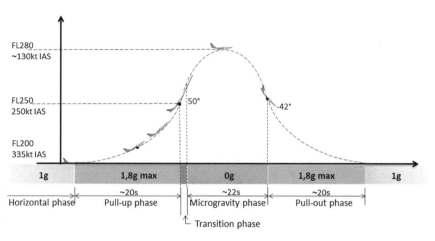

Fig. 11 Flight profile of the Airbus A310 ZERO-G during a parabolic flight: (1) starting from steady horizontal flight at an altitude of approximately 6100 m, the aircraft climbs for about 20 s with accelerations between 1.8 and 2 g; this phase is called the pull-up; (2) when the climbing angle reaches 50° at an altitude of approximately 7600 m, the pilots strongly reduce all engine thrust for about 20–25 s, keeping enough thrust to balance the air drag; the sum of all forces other than gravity is null, the aircraft is in a parabolic ballistic free fall; (3) when the diving angle reaches 42° below the horizontal, it is time for the pilots to increase again all engine thrust to dive downward accelerating around 1.8 g for approximately 20 s, to come back to a steady horizontal flight; this phase is called the pull-out. (Credit: Novespace)

A football or a rugby ball thrown by hand with an angle of 45° and an initial velocity of 20 m/s travels a horizontal distance of approximately 40 m. The Airbus A310 'ZERO-G' (see Fig. 11) initiating a parabolic trajectory with an initial velocity of 500 km/h inclined by 50° on the horizontal describes a parabola of 21.7 s duration, travels a horizontal distance of 1936 m and climbs approximately 580 m above the altitude of the parabola injection.

One can also verify that this parabola is actually an arc of ellipse, but the approximation committed is so small (in the order of 10^{-4}, see Ref. 25) that it is perfectly acceptable.

If one throws upward a body from ground with a vertical initial velocity v_0 ($\theta = 90°$), the maximum height reached is

$$h_{max} = \frac{v_0^2}{2g} \tag{A7.9}$$

before falling back to Earth. However, beyond a certain initial velocity, the body does not fall back anymore on Earth. The limit initial velocity beyond which the body escape from the planet gravitational attraction is called escape velocity (or first cosmic velocity in the Russian literature).

It can be calculated easily by finding the work required to fight against the gravitational attraction and kinetic and potential energies.

To move against the force of gravity a body initially on Earth's surface of an elementary vertical distance dr, one must expand an elementary work dW

$$dW = F_{gr}dr = \frac{G.M_E.m}{r^2}dr \tag{A7.10}$$

that is transformed in potential energy. The maximum potential energy is equivalent to the total work expanded to bring the body from Earth's surface to an infinite distance for the body to escape terrestrial gravitational attraction (if one neglects the influence of all other masses in the Universe):

$$E_{pot} = W_{pot} = \int_{r_E}^{\infty} F_{gr}dr = \int_{r_E}^{\infty} \frac{G.M_E.m}{r^2}dr = \frac{G.M_E.m}{r_E} \tag{A7.11}$$

On the other hand, a velocity v induces a kinetic energy in a body of mass m

$$E_{cin} = \frac{m.v^2}{2} \tag{A7.12}$$

The escape velocity v_{esc} is the initial velocity necessary to give the kinetic energy to the body to attain the state of maximum potential energy. In other words, the escape velocity must give the body a sufficiently large kinetic energy to balance the potential energy due to Earth's gravitational attraction.

If one neglects the dissipation of kinetic energy in heat by friction in the atmosphere, this velocity is calculated by the total energy conservation between the initial state (where kinetic and potential energies are respectively maximum and null) and the final state (where kinetic and potential energies are respectively null and maximum), yielding:

$$\frac{m.v_{esc}^2}{2} = \frac{G.M_E.m}{r_E} \tag{A7.13}$$

or with relation (5) $g_{gr} = \frac{G.M_E}{r^2}$ (see Chapter 1) of gravity acceleration at Earth's surface

$$v_{esc} = \sqrt{\frac{2G.M_E}{r_E}} = \sqrt{2g_{gr}r_E} \tag{A7.14}$$

At Earth's surface, the escape velocity is

$$v_{esc\,E} = 11186 \text{ m/s} \approx 11.2 \text{ km/s} \tag{A7.15}$$

The interplanetary spacecraft have initial velocities greater than this value when leaving the low Earth orbit.

Appendix 8: Perturbations Inside a Satellite

Let us consider the ideal case of a satellite having a spherical shape and mass distribution, such as its inertia centre is at the satellite geometrical centre. In its orbital movement around the Earth, it is the inertia centre that describes the circular or elliptical free fall orbital trajectory. An object inside the satellite located at a certain distance from the satellite's inertia centre is subject to a small perturbation force due to this distance. The order of magnitude of this force can be calculated with the following hypotheses:

- Earth is spherical of mass M_E;
- the satellite has a mass m_S negligible compared to Earth's mass;
- the satellite orbit is circular and external perturbation forces (such as e.g. atmospheric drag) are negligible;
- the satellite presents always the same side to Earth, i.e. the satellite performs a complete rotation on itself while it describes a revolution around Earth (the two angular velocity vectors of orbital revolution and satellite's rotation are parallel and have the same magnitude).

One considers the two following reference frames (see Fig. 12):

- a celestial geocentric reference frame R_E whose origin is at Earths' centre O and that one supposes inertial, in which the satellite movement is expressed;
- a reference frame R_S whose origin is at the satellite inertia centre C and axes are fixed with respect to the satellite, one of them being along the radius vector from

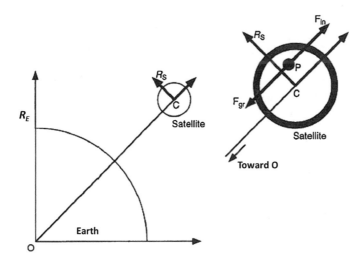

Fig. 12 Forces applied to an object located in P away from the inertia centre C of a spherical satellite on a circular orbit around a spherical Earth. The celestial geocentric reference frame R_E is inertial; the reference frame R_S attached to the satellite inertia centre is non-inertial

Earth's to satellite's centres. As the satellite always shows the same side to Earth, in its orbital movement, the satellite remains immobile in this reference frame. This reference frame R_S is non-inertial as it is in rotation with respect to the geocentric reference frame R_E and one describes the movement of an object in the satellite with respect to this reference frame R_S.

In the inertial reference frame R_E, the only force that acts on the satellite is Earth's gravitational attraction; there are no other forces, in particular no centrifugal inertia force. In the satellite reference frame R_S, two forces act on all points of the satellite, in particular on the satellite inertia centre C:

(1) the gravity force due to Earth gravitational attraction

$$\vec{F}_{gr} = -\frac{G.M_E.m_S}{r_{oc}^2} \cdot \left(\frac{\vec{r}_{oc}}{r_{oc}}\right) \tag{A8.1}$$

(2) the centrifugal inertia force due to the satellite orbital movement

$$\vec{F}_{in} = m_S.\omega_{orb}^2.\vec{r}_{oc} \tag{A8.2}$$

where ω_{orb} is the satellite orbital angular velocity.

These two forces are equal in magnitude but of opposite direction. One deduces easily the expression of the orbital angular velocity:

$$\omega_{orb} = \sqrt{\frac{G.M_E}{r_{oc}^3}} \tag{A8.3}$$

Consider an object of mass m_p released without initial velocity from a point P inside the satellite and in the same orbital plane as the satellite inertia centre C. In the inertial reference frame R_E, the only force that acts on the object is Earth's gravitational attraction (one neglects the attraction due to the satellite mass) and there is no centrifugal inertia force. In the reference frame R_S attached to the satellite, the following forces are acting:

(1) Earth's gravitational force

$$\vec{F}_{gr} = -\frac{G.M_E.m_p}{r_{op}^2} \cdot \left(\frac{\vec{r}_{op}}{r_{op}}\right) \tag{A8.4}$$

(2) the centrifugal inertia force due to two rotation movements, the whole satellite orbital revolution movement and the satellite rotation on itself. This force reads

$$\vec{F}_{in} = m_p \cdot \omega_{orb}^2 \cdot \vec{r}_{oc} + m_p \cdot \omega_{rot}^2 \cdot \vec{r}_{pc} \tag{A8.5}$$

As the satellite always shows the same side to Earth, its rotation velocity is equal to the satellite orbital velocity:

$$\omega_{rot} = \omega_{orb} \qquad (A8.6)$$

In the reference frame R_S attached to the satellite, the resultant of the two forces given in (A8.4) and (A8.5) is no longer exactly null and a small perturbing inertia force $\vec{F_p}$ exists due to the distance CP of the object from the satellite inertia centre. The norm of this force is with a good approximation

$$F_p \approx 3 \frac{G.M_E.m_p}{r_{oc}^3}.r_{cp} = 3 \frac{G.M_E.m_p}{r_{oc}^2} \cdot \left(\frac{r_{cp}}{r_{oc}} \right) \qquad (A8.7)$$

It can be understood as the gravity force applied to the object mass m_p placed at the satellite inertia centre C, corrected by a factor equals to three times the ratio of the distance r_{cp} of the object in P to the satellite inertia centre C and the distance r_{oc} of this inertia centre C to Earth's centre. The object, as soon released from a point connected to the satellite (therefore moving before the release at the satellite orbital angular velocity ω_{orb}) starts to describe its own orbit around Earth inside the satellite. Seen from the satellite inertia centre, the object drifts slowly until it touches a satellite wall.

For an orbit at 500 km altitude, an object inside a satellite is subjected to a parasitic acceleration of 4×10^{-6} m/s^2 per metre of distance.

The cases of non-rotating non-spherical satellites are treated similarly, but with a more complicated formalism. The magnitude orders of perturbations however are similar.

Appendix 9: Incorrect Ideas About Weightlessness

We correct here some incorrect or approximative ideas on weightlessness and its causes, but that are unfortunately very common and widespread. The informed reader will understand that these clarifications are necessary. Indeed, it is important to not only say what weightlessness is, but also what it is not. Experience shows that some of these incorrect ideas die hard.

1 Weightlessness is not Due to the Nulling or Disappearance of Gravitation

One never insists enough on this point: the disappearance of gravitation or gravity is impossible. The only theoretical cases (fortunately!) are the simultaneous nulling of all masses in the universe, or being at an infinite distance of all masses in the Universe, both being physically meaningless.

2 Weightlessness is not Due to Absence of Atmosphere

Evacuating a closed volume would not mean that a proof mass released in this volume would not fall anymore. To the contrary, it would fall even more easily and slightly faster as there is no longer air friction.

3 Weightlessness is not Due to a Large Distance Away from Earth or Any Other Mass

Weightlessness is currently generated on Earth. It is sufficient to put oneself in a state of free fall. Jump two steps from the staircase and, for a fraction of seconds, you are in a state of weightlessness (if air friction is neglected). Being away from all masses in the universe is of course impossible. This incorrect idea is close to the one of nulling gravitation: "if you are far enough from all attractive masses, you do not feel the attraction anymore or very little". This is wrong of course: the diminishing of the force of gravity does not cause weightlessness. To the contrary, it is because the masses in the Universe that gravitation exists and that free fall can exist. The state of free fall is a dynamic phenomenon: the body must be in motion in a gravitational field. It does not help to try to get away from gravity. To the contrary, one must abandon oneself to it.

4 Weightlessness is not Due to a Very High Velocity

This idea is probably caused by an analogy with an airplane at take-off that must reach a certain speed to be airborne and to be able to fly freely in the air. If it is true that sustentation or lift of an airplane is a function of velocity by Bernoulli's principle, it has nothing to do with weightlessness. Moreover, the high orbital velocities of satellites in orbit around Earth (approximatively 28 000 km/h or 8 km/s) can contribute to the confusion. Let us recall that weightlessness is immediately achieved even with a null initial velocity. An object is directly in free fall as soon as it is released from the hand and therefore, in a weightless state in its own reference frame.

5 Weightlessness is not Only Obtained at a Point of Equal Attraction Between Two Masses

This idea very unfortunately was popularized by the French Author Jules Verne in his book "Around the Moon". He describes weightlessness as the state that only lasts the instant of passage to the point of equal gravitational attraction between Earth and the Moon. If it is exact that weightlessness does exist when passing by this point (if there are no other non-gravitational forces), it also exists at every moment of the unpropelled ballistic flight between Earth and Moon, as demonstrated by the Apollo Moon missions.

6 Weightlessness is not Simply Due to a Balance Between Gravity and Centrifugal Forces

Here, the situation is more delicate and one must be careful. The statement, often made as is, of the simple balance between two forces, one directed "inside" the movement, the other one directed "outward" of the movement, is incorrect because incomplete. The reference frame in which this statement is made is not specified and it makes it wrong. In a reference frame attached to the body in free fall, the statement is correct as the reference frame is non-inertial: inertia forces therefore exist in this reference frame to balance the gravity force. In an external reference frame, that can be considered as inertial most of the time, the statement is wrong because there are no inertia forces, and the body only falls freely due solely to the force of gravity (if there are no perturbing forces).

Let us mention as well that there is no way to recreate weightlessness in centrifuges (by making them turn "the other way around" for example). Weightlessness can be simulated on Earth by certain methods that will be described in Chapter 3. Weightlessness can also be created on ground, but always following the principle of free fall, as it will be explained in Chapter 3.

Chapter 3
Weighing in on Microgravity?

Introduction to the Third Chapter

In the first chapter, the notions of inertia forces and inertial and non-inertial reference frames (respectively without and with inertia forces) were introduced. *Gravity* (weightiness) was defined as the phenomenon resulting from gravitation and inertia. In the second chapter, we saw that a body submitted to the only force of gravity is in free fall in an external inertial reference frame and in weightlessness in its own non-inertial reference frame.

In this chapter, microgravity is defined in Sect. 7. Section 8 presents methods and means used to create microgravity are presented, and the advantages of this environment are described in Sect. 9, with an example of application addressed in Appendix.

7 Microgravity

Practically, perfect weightlessness is impossible to realize in a free-falling vehicle. The reasons for this are simple in their principle but complex to evaluate. Simply said, there always remain some little parasitic forces creating residual accelerations. These parasitic forces can be classified in two categories: the non-gravitational forces "external" to the vehicle, whose resultant is not null; and the "internal" forces for which inertia forces do not compensate exactly gravity. Appendix 8 presents some of these perturbing forces.

The terms of weightlessness, null *gravity* (weightiness) or zero *gravity* (weightiness) designate an idealized state that cannot be obtained. One talks about reduced *gravity* (weightiness) or micro-*gravity* (weightiness), to be taken in its etymological sense of the prefix micro- (from Greek *mikros*, small) and not in its usual technical sense, millionth. As already indicated previously, the language abuse consisting of replacing *gravity* (weightiness) by gravity unfortunately is widespread. One often

© The Author(s) 2018
V. Pletser, *Gravity, Weight and Their Absence*, SpringerBriefs in Physics,
https://doi.org/10.1007/978-981-10-8696-0_3

finds the terms of null gravity, zero gravity, reduced gravity and microgravity. This last term is now consecrated to designate micro-*gravity* (weightiness). To follow the usage common to the domain specialists, we will from now onwards commit the language abuse of talking about gravity (and its derivative, microgravity and hypergravity) while it is *gravity* (weightiness) that we really mean.

Note as well that microgravity does not refer to the gravitational attraction of another body than Earth, having a small mass with respect to Earth's, but close to the location considered in microgravity, e.g. the mass of an orbital vehicle in which an experiment is conducted.

One specifies the "quality" of the obtained microgravity by the magnitude of the residual acceleration, most commonly expressed as a fraction of g_w, the average acceleration of *gravity* (weightiness) on Earth (the subscript w in g_w will be dropped from now onwards). In the next section, we describe the methods used to create microgravity. To fix orders of magnitude, one obtains residual accelerations in the order of 10^{-6} g (a millionth of the *gravity* acceleration) in drop towers and from 10^{-2} to 10^{-4} g (a hundredth to a ten thousandth of g) in the International Space Station (ISS) in orbit between 400 and 450 km altitude.

Other important factors quantify the "quality" of obtained microgravity.

Firstly, a parameter that intervenes as soon as one talks about accelerations is the frequency with which residual accelerations are applied i.e. in case the acceleration is not constant and intermittent only, the frequency is the number of times the acceleration changes per second. One specifies the acceptable level of residual acceleration in fraction of g in function of frequency ranges to which the observed phenomena are sensitive. One tries as far as possible to eliminate too important residual accelerations by isolating experiment equipment. Generally speaking, most of experiments are less sensitive to high frequencies (above 10 Hz) than to low frequencies (less than 1 Hz).

Another parameter characterizing microgravity is the duration during which the state of quasi-weightlessness is established; it depends on the chosen vehicle and on the environment in which it falls freely. Duration goes from a few seconds in drop towers to several years on board a space station.

The available volume for the experiment equipment and the type of interaction to make it functioning are two other important parameters. One distinguishes between an indirect interaction mode, either automatic or remotely operated, and a direct interaction mode by on board human operators. Furthermore, data can be either recorded on board and transmitted later, or directly transmitted allowing a quasi-immediate remote interaction.

The last important parameter is the economic aspect: the cubic metre of experiment costs more to launch in orbit than to fly in a laboratory-aircraft.

8 The Means of Microgravity Creation

All the means to create microgravity are based on the principle of free fall; any other method will not result in a real microgravity environment but in a simulated microgravity environment. Microgravity is created in a non-inertial reference frame attached to a vehicle in free fall, on which the resultant of forces other than gravity is null or negligible. The means of microgravity creation are defined here after in function of the described trajectories.

1 Vertical Free Fall Trajectories

Vertical free fall trajectories are obtained in free fall facilities. There are three types of facilities: drop tubes, drop towers and atmospheric capsules.

A drop tube is a vertical tube in which a capsule containing experimental equipment can fall freely. Three systems can be used to overcome air resistance or atmospheric drag:

- either evacuate the tube or tower;
- or slightly accelerate downward the capsule by a propulsion rocket;
- or protect the capsule by a shield placed at a certain distance from the capsule at the beginning of the fall and released simultaneously with the capsule.

The level of microgravity obtained in the drop tube of NASA Marshall Centre of 105 m high and 25 cm diameter, is in the order of 10^{-6} g during 4.6 s in a vacuum of 10^{-9} bar (a billionth of the normal atmospheric pressure).

Drop towers are used for more massive capsules. In Europe, the ZARM drop tower in Bremen, Germany, (see Fig. 13) is 110 m high with a diameter of 3.5 m.

In this tower, experimental equipment up to 170 kg (see Fig. 14), falling during 4.7 s in a vacuum of 10^{-5} bar, can attain microgravity levels of 10^{-5} g. The microgravity duration can be doubled up to 9.5 s by launching the experiment capsule in a catapult mode from the bottom of the tower upward, falling freely first upward and then downward.

The National Microgravity Laboratory of the Chinese Academy of Sciences uses since 2003 an 83 m high drop tower in Beijing (see Fig. 15). Two kinds of capsules are used, either a single drop capsule or a double capsule, the outer one acting as a shield for an inner capsule in vacuum. These capsules can fall freely for 61 m providing 3.5 s of microgravity levels of respectively 10^{-3} g and 10^{-4} g.

Despite the very low residual acceleration levels, limitations in size and duration render the use of drop towers and tubes interesting only for specific applications or investigations of short duration phenomena.

Fig. 13 The ZARM Drop Tower in Bremen, Germany. The 146 m high building protects the free fall facility from atmospheric perturbation and wind. Microgravity levels of 10^{-5} g are obtained during 4.7 s in drop mode and 9.5 s in catapult mode. (Photo: ZARM)

Fig. 14 The experimental capsule of 81 cm diameter and 1.85 m high, lifted to the ZARM tower top. The capsule aerodynamic profile and the fall in vacuum of 10^{-5} bar allow to minimize to the extreme the atmospheric friction. The decelerating container of 8 m high, filled with polystyrene balls, is seen to the left. (Photo: ZARM)

The third type of free fall facility is a system that was developed in Germany, under the name of "*Mikroba*", where the experimental capsule shaped as a rocket is

Fig. 15 The Drop Tower of the National Microgravity Laboratory of the Chinese Academy of Sciences in Beijing, China, provides a 61 m free fall yielding 3.5 s of 10^{-3}–10^{-4} g environment. (Photo: NML-CAS)

uplifted by an atmospheric balloon up to an altitude of several tens of kilometres and released remotely. A rocket engine gives a downward propulsion increasing with the density of the atmosphere layers encountered during the fall, yielding a level of 10^{-3} g during 60 s. Although 17 km are sufficient for a free fall of 60 s, a large part of the trajectory is used for deceleration. Furthermore, it is important that free falls take place in an as quiet as possible atmosphere. This system however was abandoned more than twenty years ago for economic reasons.

2   Parabolic Free Fall Trajectories

Parabolic free fall trajectories are described by laboratory aircraft and sounding rockets.

Aircraft parabolic flights allow to obtain microgravity levels in the order of 10^{-2}–10^{-3} g, with the additional advantage of having on board human operators who can intervene directly in the experiment performance during the microgravity phase (see Fig. 16).

The volume in which the microgravity is created for performing experiments is as large as the aircraft cabin.

The microgravity duration depends on the initial velocity of the aircraft at the beginning of the ballistic phase. Typical duration values are between 20 and 25 s for the NASA DC-9, the Russian Ilyushin IL-76 MDK and the CNES/ESA/DLR Airbus A310 ZERO-G (see Fig. 17) and previously NASA KC-135 and the CNES/ESA Airbus A300 and Caravelle airplanes.

Sounding rockets flights, for which microgravity levels are in the order of 10^{-4}–10^{-5} g, are used for automated or remotely operated experiments with relatively reduced volumes. There are four types of sounding rockets used in Europe (see Fig. 18): Mini-TEXUS, TEXUS, MASER and MAXUS, and yield microgravity durations between 3 and 13 min. The Programme MAXUS has been presently

Fig. 16 During a parabolic flight on board the Airbus A300 ZERO-G during an ESA campaign, several experimental racks are visible to the left and the back while the author floats freely "upside down". These is no "up" and "down" in weightlessness. (Photo: ESA)

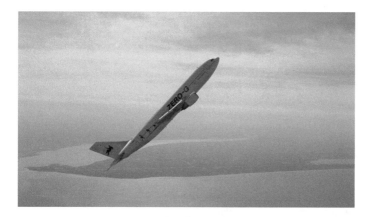

Fig. 17 The Airbus A310 ZERO-G during the pull-up. (Photo: Novespace—Eric Magnan/Airborne Films)

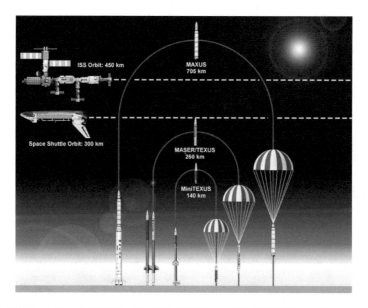

Fig. 18 Flight profiles of MAXUS, MASER, TEXUS and Mini-TEXUS sounding rockets used for microgravity experiments. Launch takes place typically from Kiruna in North of Sweden. The propelled flight lasts several minutes up to shown altitudes (higher than ISS for MAXUS) until separation of the propulsion and the experiment carrying modules. The ballistic phase above Earth's atmosphere lasts typically 3 min for Mini-TEXUS, 6 min for TEXUS and MASER and 13 min for MAXUS. Experiment performance is followed from ground by data and video signals telemetry. During reentry, the experiment stage is braked by a parachute. After soft landing, the payload is recovered and returned by helicopter within the hour to scientists at the launch base. (Credit: ESA)

Fig. 19 Three suborbital facilities in development: (left) the New Shephard capsule with a reusable rocket (Credit: Blue Origin); (middle) the WhiteKnightTwo airplane carrying the SpaceShipTwo spaceplane (Photo: Virgin Galactic); (right) the Lynx spaceplane (Credit: XCOR)

discontinued for economic reasons. The REXUS and Mapheus sounding rockets, similar to Mini-TEXUS, are used for student experiments and provide microgravity for 3 min.

In the USA, several programs of sounding rockets exist, some for microgravity research and other for atmospheric sounding, astrophysics or general physics research. The Spaceloft XL, launched from Southern New Mexico, provides microgravity for 4 min. RockSat-X is a programme for students using NASA Terrier-Malemute rockets. In Brazil, the VSB-30 rocket provides microgravity for 6 min. The Australian Space Research Institute Ltd. uses Zuni rockets for educational investigations with microgravity duration of 2–3 min.

In a near future, suborbital flights will provide microgravity duration in the order of 3 min for paying customers, but also for microgravity experiments. There are typically three US companies that are working on suborbital vehicles (see Fig. 19): Blue Origin with the New Shephard capsule and a reusable rocket; Virgin Galactic and the SpaceShipTwo spaceplane carried by the WhiteKnightTwo airplane carrier; and XCOR Aerospace with its Lynx spaceplane (this programme is presently on hold for an undefined duration).

These three systems would carry passengers and experiments up to an altitude of 100 km in a propelled mode and continue in a ballistic mode for approximately 3 min after propulsion has stopped.

3   Circular and Elliptic Orbital Free Fall Trajectories

Circular and elliptic trajectories are realized in orbit around the Earth. Manned orbital platforms provide microgravity periods of several years for the International Space Station (ISS), with its first element launched in 1998 and assembly completed in 2011 (see Fig. 20), and the future Chinese Space Station foreseen to be assembled in orbit in 2022 (see Fig. 21). Residual accelerations are in the order of 10^{-2}–10^{-4} g, depending on internal perturbations (e.g. crew movements, see Fig. 22) and external ones (non-gravitational forces, see Appendix 10).

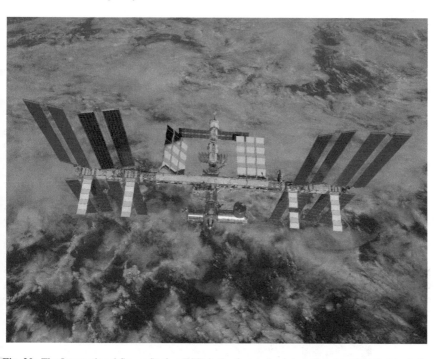

Fig. 20 The International Space Station (ISS) is the first major international project that includes 14 countries in its realization: USA, Russia, Canada, Japan and ten European countries: France, Germany, Italy, Belgium, The Netherlands, Spain, Sweden, Switzerland, Denmark and Norway. Other countries have also participated with some experiments on the ISS. With a total mass of 440 Tons (but weighing 0 kg…), the ISS is in low-Earth orbit between 400 and 450 km altitude at 51.6° inclination. Several vehicles were used to assemble it between 1998 and 2011: NASA Space Shuttle, Russian Proton launchers with Soyuz and Progress vehicles, Ariane 5 rockets with ESA Automated Transfer Vehicle, Japanese H2 rockets, and commercial Space-X Dragon and Orbital Cygnus vehicles. Simply said, the ISS consists of three parts: (1) the structure: a central beam of 100 m long supporting 16 solar panels of approximately 40 m long each and a series of accordion-like radiators; (2) the modules: from top, two Russian modules, *Zarya* and *Zvezda* (meaning respectively Dawn and Star) connected to a Node 1 called *Unity*, to which is connected the US Laboratory *Destiny*, to which is attached the Node 2 *Harmony*, connecting the European *Columbus* Laboratory (to the left) and the Japanese module *Kibo* (meaning Hope, to the right); (3) the vehicles: *Soyuz*, carrying astronauts and cosmonauts, and various automated cargo vehicles carrying resources (food, water, instruments, tools, …). Since November 2000, the Station is inhabited by permanent international crews. (Photo: NASA)

Automatic orbital platforms, as e.g. Russian satellites *Foton* (see Fig. 23) and *Bion*, and Chinese satellites of the *Shin Jian* series (meaning "Practice" in Chinese Mandarin, see Fig. 24), provide excellent microgravity levels in the order of 10^{-5} g for several weeks or months.

Fig. 21 The Chinese Space Station foreseen to be launched and assembled in the coming years with an assembly completed for 2022. From left, the *Tianzhou* (meaning "Heavenly Vessel" in Chinese Mandarin) cargo freighter docked to the *Tianhe* ("Harmony of the Heavens") core module in the centre; a piloted *Shenzhou* ("Divine Vessel") vehicle is connected to a node in front of *Tianhe*, to which are connected two scientific modules, *Wentian* ("Quest for the Heavens", at right) and *Mengtian* ("Dreaming of the Heavens", left). (Credit: CMSA)

Fig. 22 ESA astronaut Luca Parmitano installing the FASES experiment in the Fluid Science Laboratory in the European Columbus Laboratory on the ISS in June 2013. (Photo: ESA)

Fig. 23 On a low Earth orbit at approximately 300 km altitude and 62.8° inclination for 12 days in September 2007, the *Foton* M3 capsule carried a 400 kg European experiment payload with experiments in fluid physics, biology, crystal growth, radiation exposure and exobiology. (Credit: ESA)

Fig. 24 The Chinese *Shi Jian*-10 ("Practice") microgravity science satellite allowed to perform twenty microgravity experiments during a 13-day mission in orbit in April 2016, after which the Entry Module of the satellite separated and re-entered the atmosphere to be recovered on ground, while the Orbital Module continued orbiting until July 2016 to complete fluid physics experiments that did not need to be recovered. With a total mass of 3 600 kg, including a 600 kg payload mass, the *Shi Jian*-10 spacecraft orbited Earth at approximately 350 km with an orbit inclination of 42.9°. (Credit: NSSC/CAS)

4   Comparison of Different Means

Space missions of orbital platforms and of sounding rockets require a long preparation, typically of several years and should be considered for experiments that need a long exposition duration to microgravity (see Fig. 25). The relatively short preparation time for the use of drop tubes and towers and of aircraft parabolic flights (typically of few days to few months) render them particularly attractive for short duration experiments of a few seconds. The utilization of these experimental platforms of earthbound microgravity must be considered as preparatory and complementary to space missions.

Table 3 and Fig. 26 summarize the comparison between the different vehicles in which microgravity is created. In Table 3, the working volume refers to an average volume for experimental equipment and indirect human interaction implies the use of telecommands.

Let us insist on the fact that these means do not simulate microgravity, but that they really create microgravity, even if it is not always perfect, as all these means are in free fall. Appendix 11 gives several methods used to simulate a microgravity environment, but without really creating it.

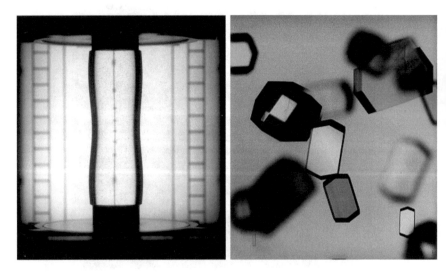

Fig. 25   Two typical examples of experiments needing several hours exposition to microgravity. (Left) A fluid physics experiment performed with the Advanced Fluid Physics Module during the Spacelab D2 mission in 1993, during which a liquid column of silicon oil was formed between two circular discs of 3 cm diameter, one disc moving backward from an initial position close to the other feeding disc from which the silicon oil was injected in the disc centre. The total length column is 10 cm, close to the rupture theoretical limit of π times diameter (Photo: DLR). (Right) Protein crystals obtained with ESA's Advanced Protein Crystallization Facility during the Life and Microgravity Spacelab mission on NASA Space Shuttle STS-98 in May 1995. (Credit: Prof. Martial, University of Liege, Belgium)

Table 3 Typical characteristics of microgravity platforms

	Microgravity level (g)	Duration	Working volume (m^3)	Human interaction	Preparation time
Drop tubes and towers	10^{-5}–10^{-6}	Max. 5 s	<1	Indirect	Few days
Aircraft parabolic flights	10^{-2}–10^{-3}	20–25 s	>10	Direct	Few months
Sounding rockets	10^{-4}–10^{-5}	5–7 min	<1	Indirect	1–2 years
Orbital stations					
Manned	10^{-2}–10^{-4}	Several years	>1	Direct	Several years
Automatic	$\approx 10^{-5}$	Several months	>1	Indirect	Several years

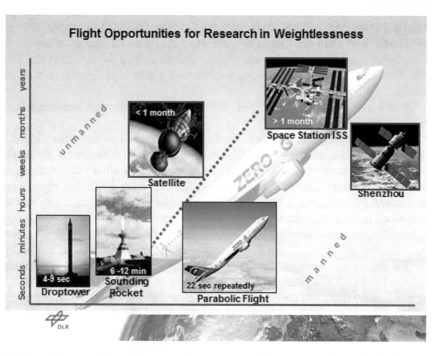

Fig. 26 Reduced gravity platforms accessible to microgravity researchers. (Credit: DLR)

9   Interest of Microgravity

The Earth's environment is characterized by the constant presence of gravity and most of the phenomena studied on Earth are subjected to its action. This environment is so much normal that one calls "natural" phenomena ensuing from gravity action. For example, convection (the movement) of fluids, liquids or gases, caused by the upward displacement of hotter zones or zones of lower concentration (therefore less dense) and the downward displacement of cooler zones or zones with higher concentration (therefore denser), is called "natural" convection.

Although physical and biological processes are often investigated in hypergravity, e.g. in centrifuge, one knows less what happens in reduced gravity. However, in most cases, one cannot extrapolate from results obtained in hypergravity to microgravity, most of the phenomena being nonlinear in function of the gravity level. One observes many more differences while passing from 1 to 0 g than between 5 and 4 g for example.

Many scientific fields profit from the peculiarities of weightlessness to enlarge their field of investigations. Material sciences, fluid physics and life sciences (biology and physiology) were the first to use microgravity, followed later by many other disciplines (combustion physico-chemistry, crystallography, fundamental physics, critical point phenomena, etc.) in view of varying a new experimental parameter: gravity.

Table 4 shows some of the scientific fields in which experiments were conducted in microgravity.

Microgravity research allows to study the gravity effects on these different phenomena and the effects of other forces normally masked by gravity on Earth. Weightlessness became an experimental research tool that allows to transpose in microgravity the investigation of phenomenon known on Earth but sometimes insufficiently understood, in order to investigate the fundamental processes and to understand their functioning without gravity.

Modifications appear when one studies matter behaviour in weightlessness. One observes on one hand the disappearance of "natural" phenomena caused by gravity, and on the other hand, the preponderance in microgravity of phenomena that can hardly be observed in normal conditions of gravity. These modifications are particularly important for certain physical, chemical and metallurgical processes having at least one fluid phase: crystal growth, alloy solidification, separation of biological substances, etc.

The main differences that are observed for fluid phases in weightlessness are as follows.

Table 4 Non-exhaustive list of research fields in microgravity

Physical sciences	Life sciences
Fundamental physics	*Human research*
Complex plasmas and dust particle physics	Integrated physiology
Aerosol particle motion	Cardiovascular function
Frictional interaction of dust and gas	Respiratory function
Plasma physics	Body fluid shift
Aggregation phenomena	Central venous pressure system
	Digestive system
Materials science	Muscle and bone physiology
Thermophysical properties	Skeletal system
Thermophysical properties of melts	Blood lactate studies
New materials, products and processes	Body mass tests
Morphological stability and Microstructures	Human locomotion
Physical chemistry	Posture
Aggregation phenomena	Bone models
Granular matter	Neuroscience
	Neurobiology
Fluid and combustion physics	Vestibular functions
Structure and dynamics multiphase systems	Spatial orientation
Pool boiling	Motion sickness
Heat and mass transfer	Motor skills
Dynamics of drops and bubbles	
Thermophysical properties	*Biology*
Interfacial phenomena	Plant physiology
Dynamics and stability of fluids	Statolith movement
Evaporation	Gravitropism
Complex dynamic systems	Gravireceptors
Diffusion	Cell and developmental biology
Foams	Animal physiology
Chemo-hydrodynamic pattern formation	Ageing processes
Combustion	Electrophysiological and morphological
Droplet and spray combustion	properties of human cells
Soot concentration	Osteoblast cells
Combustion synthesis	
Laminar diffusion flames	*Technology*
Fuel droplet evaporation	ISS Experiment validation
Ignition behaviour	Phase separation technologies for biological
	fluids
Technology	Crew foot restraint
ISS Experiment validation	Crew exercise devices
Metal halide lamps	Urine monitoring system
Micro-acceleration measurement	

1 Disappearance of Separation Phenomena

Separation phenomena observed on Earth in multi-phase systems that includes a fluid phase disappear in microgravity. Sedimentation (precipitation of dissolved or suspended matter) and Archimedean buoyant force (or buoyancy, i.e. the force due to a liquid pressure on a body immersed volume) disappear. The advantage of absence of separation in weightlessness is with the possibility to obtain mixtures that are unstable on Earth and material alloys impossible to obtain on Earth or with great difficulty. A disadvantage of the absence of separation in weightlessness is the difficulty to eliminate the gaseous inclusions while on Earth, degassing is done "naturally" (gaseous zones in liquid matrices go up to the free surface).

2 Disappearance of "Natural" Convection

"Natural" convection disappears in fluids in microgravity. There is no more natural upward displacement of hot zones and downward displacement of cold zones. In fact, there is no up and no down. Other forces become dominant for movements in liquids in microgravity. These forces are linked to superficial or interfacial tension between two liquids. Indeed, such an interface behaves as an elastic "membrane" whose tension is a thermodynamic function of temperature (or concentration for solutions), as shown in Fig. 27.

Fig. 27 Liquid/gas interface submitted to a superficial tension gradient, yielding a Marangoni convection cell caused by the physical displacement of the interface membrane from the hot side (point 2) to the cold side (point 1)

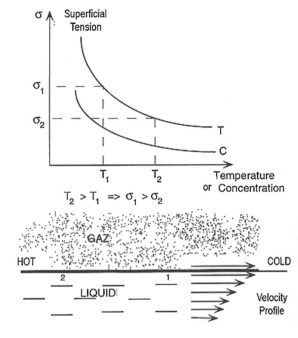

For an interface subjected to a temperature difference, superficial tension for most liquids is generally smaller for the hot side than for the cold side. The interface, i.e. the common layer formed by molecules of both fluids, physically moves parallelly to itself from the hot side to the cold side; this membrane deforms itself and slides from the hot side to the cold side. The liquid layers on both sides of the interface are dragged along by viscosity and a new convection appears, called Marangoni convection, after the name of the Italian physicist who studied this phenomenon at the end of the 19th century. This phenomenon exists obviously also on Earth, but as its effect is much smaller than those caused by gravity, it is in general negligible and much more difficult to observe. Its study in microgravity allows thus to better understand the fundamental characteristics of liquid behaviour.

Marangoni convection allows also to explain the movement of gas bubbles or liquid drops in a liquid matrix heated on one side and cooled on the other side (see Fig. 28). The curved gas/liquid (or liquid/liquid) interface between the bubble (or the

Fig. 28 Movement of a gas bubble from the hot to the cold side in a liquid matrix submitted to a temperature gradient, hot on top, cold at bottom. (Credit: ESA)

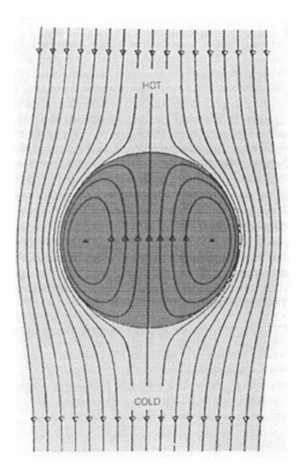

Fig. 29 Flames on ground in
1 g (left) and in microgravity
in near 0 g (right). Notice the
near-hemispherical shape of
the flame in microgravity
with the reddish-purple part
on top due to some
convection caused by small
perturbations in the
microgravity environment.
(Photo: NASA)

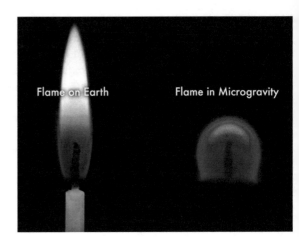

drop) and the liquid matrix behaves like the flat interface in the previous example:
it moves from the hot side toward the cold side on both sides of the bubble (or the
drop). The fluid layers inside the bubble (or the drop) are dragged along by viscosity
and are closing in a toroidal convection cell inside the bubble (or the drop). This
convection movement yields the displacement of the centre of the bubble (or the
drop) in the opposite direction than the one of the superficial layers' displacement,
i.e. from the liquid matrix cold side toward the hot side.

It is also because the absence of "natural" convection that the shape of a combustion flame is different in weightlessness. On Earth, gases produced by the chemical
reaction of combustion (of a candle wick for example), much hotter, rise and fresh air
oxygen migrate to the combustion centre to feed the combustion process. In microgravity, hot gases have no reason to rise anymore and the flame is surrounded by a
hemispherical ball formed by combustion gases (see Fig. 29), limiting the amount
of fresh oxygen transfer.

Experiments of this kind in microgravity (among others, in parabolic flights) have
shown that oxygen arrives at the flame after all, but by diffusion through the burned
gases zone.

3   Disappearance of Hydrostatic Pressure

In microgravity, hydrostatic pressure disappears. On earth, it is responsible for the
tendency of fluids to deform under the effect of their own weight, a liquid zone
supporting the weight of zones on top. The same phenomenon exists for solids.
Structures can be built that would collapse under their own weight on Earth, e.g.
crystalline networks (see Fig. 25), or metallic structure like the International Space
station (see Fig. 20).

Fig. 30 Water drop in near free float during a Chinese Shenzhou mission. (Credit: CMSA/CCTV)

Liquids in weightlessness, abandoned to themselves without any contact with a solid wall, have the particularity to form a spherical drop (see Fig. 30).

Indeed, when subjected to the only forces of superficial tension, a given liquid volume takes a spherical shape, corresponding to the minimal surface enclosing a given volume.

4 Possibility of Natural Levitation

On Earth, to melt a substance, one is obliged to use crucibles that may contaminate the melt liquid phase. In weightlessness, the liquid phase can be maintained in a contactless levitation, without touching any solid walls, using an electrostatic, magnetic or acoustic confining (see Fig. 31). Many parameters of materials at high temperatures are still unknown and cannot be measured on Earth due to difficulties and limitations caused by crucible contamination. Microgravity allows to deepen scientific knowledge in domains that are hardly accessible on Earth.

The list of advantages and applications of microgravity to scientific research could be continued at length, but is outside the aim of this publication. The interested reader will find other examples and more details in Refs. 8–11. Nevertheless, an example of practical application is given in Appendix 12.

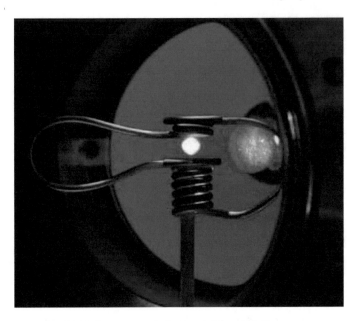

Fig. 31   Core element of an Electromagnetic Levitator. (Photo: DLR)

Conclusion of Third Chapter

Arrived at the end of this chapter, we can now answer to the question asked in the title. Yes, one should weigh in on microgravity. It allows a new experimental approach by giving the possibility to study phenomena where gravity is nullified or strongly reduced. Although known since several centuries and made easily accessible to researchers since the development of space techniques, this environment without gravity allowed since several decades to obtain an impressive amount of scientific results on phenomena intrinsic to matter or due to its state and often masked by gravity effects. Practical applications have started to appear, in particular for the study of new materials and alloys. Other practical applications concern life sciences (biology and physiology), that are the subject of Chapter 4.

Appendix 10: Perturbing Forces

Perturbing forces acting on a free-falling vehicle can be divided in two groups: "external" non-gravitational forces and forces "internal" to the vehicle.

1 External Forces

Non-gravitational forces in this case are not totally compensated. One distinguishes on one hand forces caused by phenomena external to the vehicle and thwarting its free fall movement, and on the other hand, the forces induced by the vehicle or its control systems to ensure its desired trajectory and attitude. These forces are expressed in the non-inertial reference frame attached to the vehicle inertia centre.

1.1 External Forces Caused by External Phenomena

• *Atmospheric Drag Force*

For a vehicle of mass m moving at a velocity $\overrightarrow{v}$ in an atmosphere of volumetric mass density (i.e. mass per unit volume) ρ and offering a section S projected parallelly to the movement direction, the aerodynamic drag force or the resistance of the air friction reads

$$\overrightarrow{F}_D = -\frac{1}{2}C_D \cdot \rho \cdot S \cdot v^2 \cdot \left(\frac{\overrightarrow{v}}{v}\right) \qquad \text{(A10.1)}$$

where C_D is a dimensionless coefficient whose value generally varies between 1 and 3.5, depending on the vehicle geometry and the kind of interaction between the vehicle surface and the atmospheric gas particles. The drag force direction is opposite to the velocity vector $\overrightarrow{v}$ and yields a deceleration

$$a_D = \frac{1}{2}C_D \cdot \rho \cdot \frac{S}{m} \cdot v^2 \qquad \text{(A10.2)}$$

to the vehicle. The main parameter varying with altitude is the volumetric mass ρ. For a space vehicle, like the NASA Space Shuttle, or the Russian Soyuz or the Chinese Shenzhou or automatic research satellite, this deceleration can vary between 10^{-5} and 10^{-6} g at 200 km altitude depending on attitude, and become less than 10^{-8} g at 500 km altitude, slightly above the average altitude of the International Space Station (typically 400–450 km) or of the future Chinese Space Station.

- *Radiation Pressure Force*

This force is due to the pressure exerted by electromagnetic radiations, or by the transfer of momentum from a photon flux on the vehicle surface. For a vehicle orbiting Earth, the main source is solar radiation inducing a force

$$\overrightarrow{F}_{sr} = \frac{S.A}{c}.\overrightarrow{E} \qquad (A10.3)$$

where S is the cross section on which radiations are projected (and that can be different form the section considered for the atmospheric drag), A a dimensionless coefficient varying between 1 and 2, depending on the surface thermo-optical properties, c light speed and $\overrightarrow{E}$ the total light energy flux (i.e. rate of energy transfer per unit area), whose magnitude can be correctly approximated by the solar constant (the mean solar electromagnetic radiation per unit area at a distance of 1 AU), approximately 1360 W/m^2 close to Earth. This force acts in the direction of the energy flux and yields an acceleration

$$a_{sr} = \frac{S.A}{m.c}.E \qquad (A10.4)$$

in the order of 10^{-8} g, independently from altitude.

- *Earth Infrared and Albedo Radiation Pressure Forces*

Radiation pressure forces due to Earth's own radiation, mainly in the infrared domain, and to Sun radiation reemitted by Earth, called albedo and whose effect is only sensed in the illuminated part of an orbit. These forces read as in (A10.3).

Induced accelerations are smaller by an order of magnitude than that due the direct Sun radiation (A10.4). For an orbital vehicle, these external forces slowly vary with a period close to that of the orbital revolution.

1.2 External Forces Caused by the Vehicle Control Systems

In this second category, external forces are related to vehicle manoeuvres and their variations are characterized by much shorter durations. Generally speaking, the trajectory of a vehicle inertia centre is controlled by a propulsion system and vehicle movements around its inertia centre is controlled by the attitude control system.

In case of an atmospheric vehicle (capsule in a drop tower, aircraft, rocket) in ballistic flight, propulsion force is not completely nullified but strongly reduced and can be adjusted to balance the atmospheric friction force. Attitude control is made passively by the vehicle profile (e.g. aileron or wing tip) or by differentially adjusting the propulsion system.

For an orbital vehicle, trajectory and attitude are not generally continuously controlled but sporadically, either by gas ejection of attitude engines, or by activating reaction wheels. Induced accelerations can be very important, but generally of short duration. Usually, experiments are not planned at the same time as trajectory or attitude correction manoeuvres.

2 Internal Forces

Internal forces include inertia forces that do not compensate exactly gravity. Three types of internal forces are distinguished.

2.1 Forces Due to Distance from the Inertia Centre

These forces are felt by an object that is not exactly located at the vehicle inertia centre (see Appendix 8).

2.2 Inertia Forces Due to Movement Inside the Vehicle

These forces are caused by the movement of an object in the non-inertial reference frame attached to the vehicle inertia centre. The most frequent case is the one of a satellite rotating around its inertia centre and in which an object is moving. This movement causes two inertia forces to appear, typically centrifugal and Coriolis forces (see Appendix 3 in Chapter 1).

2.3 Forces Created by the Displacement of the Inertia Centre

Let us recall that it is the centre of inertia of the entire vehicle and its content that describes the free fall trajectory, either an Earth's orbit for a space vehicle, or a parabolic ballistic trajectory for an atmospheric vehicle. The position of the inertia centre may vary in function of the variation of the mass distribution in the vehicle. This variation can be spatial (case of motion of mechanical parts) or temporal (case of fuel use). Furthermore, this variation can be slow and continuous (case of slowly moving mechanical parts) or brusque and discontinuous (case of an astronaut moving in a space station or of docking of two space vehicles).

Appendix 11: Methods of Microgravity Simulation

To the contrary of the means creating microgravity, simulation methods do not allow to really create microgravity. The simulation means allow to obtain experimental configurations in which certain aspects of phenomena can be studied in a way similar to what could be observed in microgravity but without being in weightlessness.

Therefore, these methods have important limitations that reduce their scientific interest to the investigations of some very specific cases. Results obtained by these simulation methods generally complete those obtained in real microgravity. In none of the three following configurations is microgravity really created as there is no free fall.

The first simulation method was used at the end of the 19th century by a Belgian physicist, Joseph Plateau, who gave his name to this method. The principle is simple: it consists to immerse a liquid in another immiscible liquid matrix having the same volumetric mass. By Archimedes principle, the buoyancy exerted by the liquid matrix of volumetric mass ρ_1 on a volume V of a liquid of volumetric mass ρ_2 is directed along the gravity acceleration vector and reads

$$\overrightarrow{F}_b = V. \, (\rho_1 - \rho_2) . \overrightarrow{g} \qquad\qquad (A11.1)$$

This force becomes null for $\rho_1 = \rho_2$, yielding results similar to what could be obtained in weightlessness when $g = 0$. In the Plateau configuration, the gravity force is not balanced by inertia forces but by a buoyancy force.

Only static configurations are truly well simulated with this method, e.g. configurations of static equilibrium of liquid zones. Dynamic situations of fluid movements or of thermal phenomena cannot be correctly simulated as the viscosity of liquid matrix greatly limits movement of the second liquid, and that furthermore border conditions at the two liquid interface are very different from those of free floating liquid zone in a gaseous atmosphere, due e.g. to thermal conduction at the interface.

The second simulation method is less known. It consists in balancing locally the force of gravity acting on a body by a magnetic or electrostatic force acting in the other direction. The effects of two fields, the gravitational field and a magnetic or electrostatic field, have to be locally balanced. One sees immediately the limitation of this configuration that would work only for bodies sensitive to magnetic induction or electrically charged. Furthermore, the power needed to maintain these fields is quite important and limits the size of observed configurations. Nevertheless, this method is used sometime to investigate magnetohydrodynamics problems in absence of gravity. This method is more used by Russian or eastern European researchers than by their westerner colleagues.

The third simulation method is what is called the dimensionless reduction. This method mainly applies to fluid research for which scientists use a series of dimensionless numbers describing the ratios of different forces acting on fluids. Reducing physical dimensions of an experimental liquid zone greatly diminishes effects caused by gravity in comparison to other forces acting on fluids, e.g. superficial tension force or capillarity forces. One manages to build floating liquid zone of a few millimetres size that allow to study certain phenomena. Main limitations of this method are linked to reduced sizes: firstly, they make it difficult to install precise means of observation and measurement; secondly, they reduce the field of investigation to limited ranges of values of other effects specific to fluids.

Let us also add that, for medical and physiological research on adaptation of the human body to weightlessness, researchers use two simulation techniques that allow within certain limits to recreate the effects of microgravity on the human body. It consists firstly of immobilization (or hypokinesia) in a horizontal position or slightly inclined (head down), that simulates the shift of body fluids, mainly blood, toward the upper part of the body like in weightlessness. The second technique is water immersion. As the human body is mainly made of water, buoyancy induces conditions partially similar to microgravity acting on the human body, somewhat akin to Plateau's configuration.

These microgravity simulation methods are complementary to means used to create microgravity, allowing to study in the laboratory certain aspects of phenomena appearing in weightlessness.

Appendix 12: An Example of Application of Microgravity Research

When two phases coexist (liquid/liquid or liquid/gas), fluids move along the interface by convection caused by interfacial tension gradients due to differences of temperature (or concentration) existing in the liquid matrix. This effect, called Marangoni convection, usually small on Earth in front of the gravity induced "natural" convection, depends on the temperature (or concentration) differences and is negligible in most of cases when these differences are small. For differences sufficiently large, this effect can no longer be neglected, even on Earth. Many experiments were performed in microgravity first during Spacelab missions and now on board the International Space Station to study this Marangoni effect on different liquids (solutions, semiconductors and melted alloys, …).

Monotectic alloys like Al–Bi (aluminium-bismuth) for example, are non-miscible in liquid phase. Heated above a certain temperature during sufficiently long time, the components of these alloys form a homogeneous melt. During cooling and solidification, components separate and form liquid inclusions of a component (Bi) in the other one (Al). Sedimentation and buoyancy yield eventually a complete phase separation on Earth.

It was thought that in absence of sedimentation in microgravity, this experiment on monotectic alloys would have allowed to keep the inclusions of a component (Bi) dispersed during solidification, allowing to obtain new dispersion alloys. However, it did not happen. Looking for the reasons of this failure, one realized that under the influence of temperature gradient during cooling in microgravity, liquid drops of inclusion (Bi) appearing in the melt are pushed by Marangoni effect toward hotter zones of the liquid matrix (Al). This mechanism of drop movements toward higher temperature areas due to Marangoni effect can be used on Earth to counteract sedimentation of inclusion drops caused by gravity.

Based on this principle, a new continuous casting process was conceived, employing steep temperature gradients to counteract sedimentation of droplet inclusions, with the objective of producing Al–Pb and Al–Bi monotectic alloys for self-lubricating bearings for the commercial market. The melt is poured into a mould which is cooled laterally, thus cooling the melt from the outside during solidification. The drops of lead or bismuth are propelled upwards and towards the centre by Marangoni flow, while at the same time sedimenting due to gravity. The net displacement of the drops is thus towards the centre. Homogeneous dispersions of lead or bismuth in aluminium alloys may thus be obtained, and these alloys are highly promising for applications in self-lubricating bearings.

Self-lubricating bearings are employed in all automotive engines, specifically crankshaft and camshaft bearings, where ball or roller bearings cannot be used because of the high forces that are exerted. Three regimes of lubrication are usually considered. In the first regime, during normal operation, the surface of the bearing is separated from the surface of the shaft by a film of lubricant. In the second regime, during the engine idling, the lubricating film is marginally thin, and direct point contacts between bearing and shaft material cause increased friction and wear. In the third regime, during the engine starting, the lubricant in the bearings has drained out, and bearing and shaft are in direct contact. In the second and third regimes, the self-lubricating properties of the bearing alloy prevent blocking and allow for operation at acceptable friction and wear. This is accomplished by imbedding soft metal particles, such as lead, bismuth or tin in a matrix which is sufficiently strong mechanically, as for example aluminium alloys. Aluminium alloys with dispersions of bismuth or lead had been identified as the ones with the most promising properties for self-lubricating bearings, but one has only succeeded a few years ago in producing dispersions of these alloys employing the above process.

Modern automotive engines operate at increasingly high compression and temperature, in order to increase fuel economy and reduce pollution. New materials need to be developed to meet the new requirements, and the new alloys are of this category. A comparison between the laboratory test results of standard bearing alloys and dispersion alloys obtained with this Marangoni transport effect, shows that friction is halved, and wear is reduced by a factor ten with the new alloys. The market for these new alloys is large. Several millions of cars are sold per year worldwide alone, and each four-cylinder engine needs at least eight bearings of this type.

This example illustrates the fact that fundamental microgravity research can lead to important practical applications in ways which could not have been foreseen originally. It shows the importance of maintaining fundamental research and to bring it close to applied research and engineering. More detailed information can be found in Refs. 35 and 36.

Chapter 4
Physiological Effects of Microgravity

Introduction to the Fourth Chapter

Chapters 1 and 2 of this book recalled the basic notions of inertia, inertial and non-inertial reference frames, gravitation, *gravity* (weightiness), weightlessness and free fall. One would recall that a body subjected only to the single force of gravity is in free fall in an external inertial reference frame and in weightlessness in its own non-inertial reference frame. Chapter 3 addressed microgravity, the methods used to create it, its importance and applications, mainly for material and fluid sciences.

Another category of scientific fields interested by microgravity is the whole body of researches in medicine, biology, zoology and botany, that form what is called life sciences studying the influence of gravity on living organisms, including the human being. One can distinguish two groups: disciplines concerned directly with the human body (physiology, medicine) and the other disciplines not dealing with it, or indirectly, (animal physiology and biology, cell biology, botany, radiation biology, …) and applications (pharmaceutical processes, protein crystallization, …). To review each of these disciplines would be impossible here. The interested reader can consult Refs. 8–10, 12 and 13. We limit ourselves in this fourth chapter to present some general considerations (Sect. 10) and the main physiological effects caused by microgravity (Sect. 11), looking in more details at one of the most debilitating effects, bone demineralisation (Sect. 12). This overview does not pretend to enter in all medical details (the author is not doctor in medicine nor physiologist), but to present the main lines of space medical research in microgravity. Far from being exhaustive, the text that follows targets a non-specialized public and the informed reader will forgive the non-technical language and some generalizations needed for a more general comprehension.

Recall also that the same language abuse is committed, while talking about gravity and its derivatives (microgravity, reduced gravity, hypergravity), what is really meant is *gravity* (weightiness).

© The Author(s) 2018
V. Pletser, *Gravity, Weight and Their Absence*, SpringerBriefs in Physics,
https://doi.org/10.1007/978-981-10-8696-0_4

10   General Considerations on Life Sciences Research

Space orbital environment is characterized by several factors that can influence the good functioning of all living systems, from cells to humans. The main factors are weightlessness, high energy radiations, vacuum and temperature differences. These last two factors are generally mitigated by the vehicle yielding the necessary life support to the systems under study. The first two on the contrary cannot be completely compensated. As seen previously, the state of microgravity exists in an orbital vehicle as long as it is not propelled or submitted to other important non-gravitational forces. Protection against high energy radiations cannot be actually achieved, unless thick shielding walls are installed all around spacecraft which is presently excluded in view of launch costs per kg. Nevertheless, a vehicle in low Earth orbit (a few hundred kilometres altitude) stays relatively protected by Earth's Van Allen radiation belts (inner energetic proton belt at 1 000 to 6 000 km altitude and outer energetic electron belt at 13 000 to 60 000 km altitude).

To these orbital factors, one should add the conditions at launch and during atmospheric re-entry and landing of a spacecraft, i.e. important accelerations and vibrations, that can affect quality of physiological samples or configurations obtained in microgravity (crystals for example).

Experiments performed in orbital microgravity can be spoiled by effects that can be caused by high energy radiations or by constraints at launch or re-entry. In most cases, control experiments allow to resolve certain ambiguities in obtained results.

Initially developed in the 1950s and 60s to support US and USSR space programs, space microgravity medical research quickly evolved. Manned space flights very quickly showed physiological changes in astronauts and cosmonauts. Duration of space flights have increased throughout the years, from a few hours at the beginning of the 60s to several months (or even more than a year) today on board the International Space Station (ISS, see Fig. 20). The ISS allows to conduct and to repeat experiments during several years. Numerous experiments have been conducted during Spacelab missions since the beginning of the 80s until mid-90s, and on board the Russian Mir space station since the mid-80s until end of the 90s, mainly devoted to life sciences in microgravity. New phenomena have been observed on astronauts, some of these effects appearing only after several weeks or months in space. Despite the large number of hours spent in orbit around the Earth by astronauts and cosmonauts from all countries involved in space research and exploration, some problems are still far from being fully understood and the necessary solutions have not yet been found.

Space medical research does not limit itself to conducting medical experiments in orbit, but relies also on results obtained by Earth bound means. The only suborbital mean allowing to conduct medical experiments on human subjects in real microgravity is parabolic flights (see Figs. 16 to 18). Other means of microgravity simulation, like horizontal or head-down immobilization (or hypokinesia; see Fig. 32) and pool immersion, are also used to complement microgravity medical research in orbit or during parabolic flights.

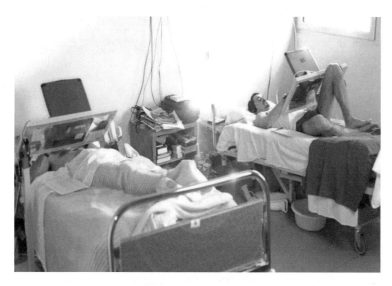

Fig. 32   Head-down bedrest simulate microgravity effects on human physiology. Subjects stay in slightly tilted head down (typically 6°) beds for weeks or months at a time (Photo: CNES/ESA)

Some phenomena observed in the human body must be confirmed by analogue experiments performed on animals or in cell biology. Interdisciplinarity of life sciences is one of the necessary characteristics of microgravity research.

Biological research fully profited of this interest. Not only, does it support space medical research, but its interest lies also in more fundamental questions. Life as we know it on Earth has evolved in a constant environment of gravity. Could it develop without gravity? The answer is not obvious. Hatching and development of fly larvae and other small animal embryos have been studied in microgravity. Results are extremely surprising to say the least. Some experiments have shown a bifurcation in the statistics of evolution of larvae and embryos populations in microgravity, half of population survives and comes to adult state, the other half dies. The reason for this dichotomy is still not understood.

11   Physiological Effects of Weightlessness

Although physiological systems of human organism function interdependently, one can classify physiological effects of microgravity in four categories:

- perturbations of sensorial systems related to balance, orientation and the vestibular system;
- modifications of bodily fluid distribution and their impact on the cardiovascular system;

- effects on metabolism and bodily functions;
- the adaptive processes of muscular and skeletal systems and their pathological consequences.

1   Balance, Orientation and Vestibular

On Earth, in a normal gravity environment, the human body has three means to obtain the information of the reference vertical direction and of the top-bottom orientation, characteristic of the gravitational environment that we know on our planet.

The main system is the vestibular system, which is double, located in the inner ear. In one of these organs, small crystals of calcium carbonate called otoliths weigh on a membrane with nervous endings (see Fig. 33).

In a static position, sitting or standing, this nervous information is transmitted to the brain that interprets it as the information giving the vertical direction. During a movement, otoliths behave as accelerometers whose activation by head movements

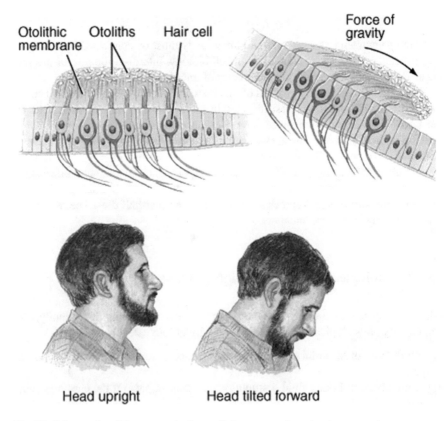

Fig. 33 Schema of otolith response in the vestibular system after a head movement

allow to transmit to the brain signals corresponding to head positions. If we move forward, inertia of these otoliths drag them backwards and the nervous information is interpreted by the brain as a forward acceleration. If we bend forward, otoliths move forward and the nervous information is interpreted by the brain as an inclination of the head. The semi-circular canals form another sensor of the vestibular system of the inner hear. Formed by three canals in planes approximatively orthogonal to each other (see Fig. 34), a physiological liquid moves by inertia in these canals during a head movement, stimulating nervous endings in the canals. The combination of these information coming from the otoliths and semi-circular canals allow the brain to interpret the movement and the position of the head.

The second source of information is the visual system. The visual information allows the brain to recognize the body position with respect to external references: floor, ceiling, walls that we know by experience are respectively down, up, and sideways. The body forward motion is also perceived by the brain through visual perception of our close surrounding moving backwards.

The third information source is the proprioceptive system, constituted of the whole of skin tactile perceptions, articulations and muscles tension. In standing position, the pressure of the body weight on the feet sole and the contact of the feet soles with the floor also give the information of the standing position to the brain. The neck proprioceptive system is the most developed and informs the brain on the position of the head with respect to the rest of the body.

In weightlessness and in absence of accelerated motion, there is no stimulation of the vestibular system. Otoliths are no longer attracted downward by gravity and the semi-circular canals are no longer stimulated. However, the visual and proprioceptive

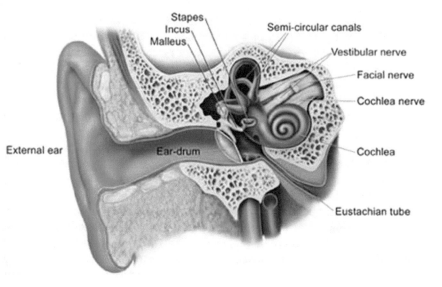

Fig. 34 View of the inner ear

systems continue to function normally. Information sent by these different systems to the brain are incoherent for an organism used to normal gravity and create confusion in the brain zone that normally treats the information on position and orientation. This confusion often yields dizzy spells and nausea, and sometime triggers the reflex of emptying the stomach. In short, the subject is sick. This sickness resembles motion sickness, although some aspects are different. This sickness, called space adaptation syndrome, or more commonly "space sickness", affects most astronauts, even the toughest ones. On average, one out of two astronauts suffers from nausea during the first few days of space flight. After a day or two, the human organism adapts to the new environment and astronauts can continue to function and work "normally".

After the flight, once back on ground, the balance and orientation system readapts very quickly to Earth's environment. This effect is therefore not the most prejudicial for the organism. Nevertheless, the loss of working capabilities and concentration of, on average, half of the crew during the first few days of flight impacts the quality and efficiency of the scientific and technical work to be done in orbit. This is taken into account by reducing the work plan and schedule during the first days

2 Body Fluids and Cardiovascular System

2.1 Loss of Body Fluids

On Earth, while standing in normal gravity, arterial blood pressure is normally distributed such that, if intracardiac (inside the heart) pressure is taken as unity, blood pressure is approximately double in feet arteries and two third at head level. While lying down, the distribution of blood pressure is more uniform. Passing from the lying to the standing position yields a blood flow toward the body lower part and blood pressure diminishes in the head. Known as orthostatic postural intolerance, the change of blood pressure is detected by baroreceptors (or pressure sensors) in the vascular system and close to the heart. These receptors send signals that yield firstly, an increase of cardiac rhythm to compensate the blood volume decrease in head arteries, and secondly, a contraction of arteries in the lower body to diminish the blood flow towards the legs.

In microgravity, gravity does not attract liquids downwards anymore and a redistribution of body fluids take place, mainly blood and interstitial liquids. A volume of approximately two litres of body fluids (approximately a litre per leg) are displaced form the lower extremities to the body upper part.

This has for consequence to increase the blood volume and pressure in the heart. The volume and blood flow receptors are alerted and this new situation is interpreted as an overload of the blood system (hypervolemia). The organism reacts as it would do on Earth, as if there was too much blood in the organism. The reaction of body liquids elimination starts and yields a complex hormonal game. First, the secretion of the anti-diuretic hormone (ADH) is diminished; this hormone is normally responsible for blocking the urinary function and helps to regulate the balance of body liquids.

This reduction of the ADH production is accompanied by the increase of secretion of another hormone, aldosterone, that regulates sodium resorption by kidneys and water retention in the organism. Other hormones intervene in this complex game, but the role of these two hormones give the global scheme of the elimination process of body liquids. Sodium being less resorbed by kidneys (water thus being less retained), associated to an increase of the urinary function, the result of an increase of blood volume is a natural elimination by urine of body liquids and sodium.

This process is similar up to a certain point and to a lower scale to what a swimmer or a diver can feel after a long stay in water. The swimmer or diver body liquids are submitted to Archimedes buoyancy due to the surrounding water and yield also a redistribution of body liquids.

The organism adapts to this new environment and a new balance is established after four to five days. Let us note that this "classical" theory is challenged by experiment results showing that body liquid redistribution should be interpreted simultaneously with other physiological effects.

On the other hand, liquid transfer from lower members toward the upper body has other secondary effects: face swelling (or oedema) due to blood rush in the head; the increase of intraocular (inside the eye globe) pressure; and sinus congestion. These secondary effects disappear up to a certain point after a few days in microgravity. Back on Earth, after flight, everything comes back to normal and the organism readapts to a terrestrial environment.

2.2 Myocardial Muscle Atrophy and Cardiac Rhythm

The decrease of the myocardium (cardiac muscle) volume was considered as a serious problem during several years. One observed indeed during the Skylab missions of the 1970s a decrease in the order of 1% of astronaut heart mass. But one should be cautious with conclusions. Indeed, after adaption to microgravity, the heart works less and an atrophy could be expected. Results of experiments performed with ultrasound echocardiography show effectively a diminution of left ventricle and auricle volumes during a space flight of several weeks. However, the initial fears of irreversibility of the cardiac muscle atrophy seem not to be founded.

One notices as well a decrease of cardiac rhythm and of arterial tension, the heart not needing to pump blood in the organism against gravity's downward pull (see Fig. 35).

On the other hand, one observed in some Russian cosmonauts an extra-systolic cardiac arrhythmia (or rhythm perturbation).

Let us note that at launch, one observes a high tachycardia, an increase of the cardiac rhythm, due to the launch psychological stress, but also necessary to compensate the effects of launch accelerations, up to maximum 8 g, eight times the acceleration of normal gravity, although the values actually are more in the order of 3 to 4 g.

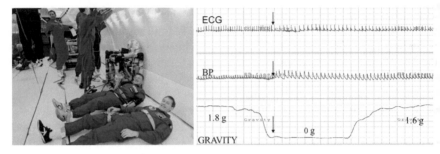

Fig. 35 Experiments during parabolic flights (left) showed (right) a decrease in heart rate, seen at the onset of 0 g (marked by arrows), i.e. an increase of duration between successive peaks, corresponding to increased vagal modulation of the heart rate. A sudden increase is also seen in pulse blood pressure (difference between maximum and minimum pressures) at the onset of 0 g indicating an increase in stroke volume. These experiments can be correlated with observations in space. (ECG: Electrocardiogram; BP: Blood Pressure) (Credit: left: ESA; right: Prof. A. Aubert, *Katholieke Universiteit Leuven*, Belgium)

2.3   Visual Impairment and Intracranial Pressure

Visual impairment and intracranial pressure is another consequence of the upward body fluid shifts, the head filling with blood and other bodily fluids. The various consequences are an increase in intracranial pressure that can cause headache of varying levels of severity, an increase of the intraocular pressure that affects the visual performance and other more minor effects such as congestion of the sinuses. These effects, although observed and investigated for several years, are thought to be temporary effects as they tend to disappear after return to Earth.

However, the first two effects, intracranial pressure and visual impairment, were only recently recognized as more serious as they could impair the performance of astronauts during long duration 0 g travels in space. Imagine an astronaut pilot who has to evaluate distances and velocities upon arriving near Mars after an eight months interplanetary journey in weightlessness.

3   *Physiological Functions and Metabolism*

In microgravity, main physiological functions are practically unchanged. Astronauts can eat, drink without major constraints. Digestion and intestinal transit are accomplished also nearly normally, except that gravity action is no longer present.

Breathing is also made without too important problems. However, the breathing mechanism is altered: the distribution of inspired and expired gases in lungs and oxygen exchanges in blood haemoglobin at the level of pulmonary alveoli are modified. The way to breathe is also modified: statistically, in weightlessness, the forced movement of the abdomen contributes more to the breathing mechanism (see Fig. 36).

Astronauts can also sleep in space. However, daily and sleep rhythms are disturbed. Indeed, on board ISS in low Earth orbit at 400 km altitude, day and night alternation repeats approximately every 90 min. Astronauts see a sunrise and sunset 16 times per terrestrial 24 h "day". Psychological and emotional factors and travel excitement intervene also. To remedy it, one imposes a strict and well-established schedule taking into account human natural rhythms. On board the ISS, a three times 8 h schedule is applied: 8 h for sleep, 8 h for work depending on missions, and 8 h for personal time, meals, rests, etc. This schedule is purely theoretical as astronauts on board ISS spend much more of their time to work, although for long duration stays on ISS, schedules are loose and longer rest periods are foreseen some days, generally used by astronauts to watch Earth through windows, mainly the cupola (see Fig. 37).

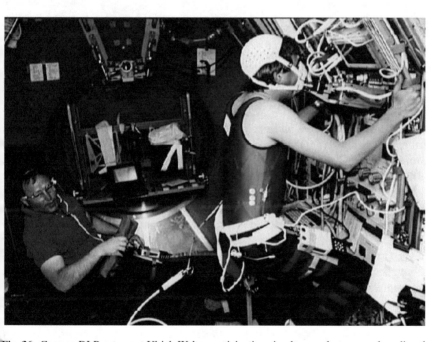

Fig. 36 German DLR astronaut Ulrich Walter participating simultaneously to several cardiopulmonary experiments using ESA Anthrorack facility during the Spacelab D2 mission: the subject is sitting and pedalling on an ergometer; he breathes through a mouthpiece of ESA's Respiratory Monitoring System, which measures the characteristics of the inspired and expired gases. He wears a Respiratory Inductive Plethysmograph jacket, which measures changes in inductance of two conducting wires (visible as zig-zag inside the blue jacket) around the rib cage and abdomen to study of the breathing mechanisms (Photo: NASA)

Fig. 37   NASA astronaut Karen Nyberg, Expedition 37 flight engineer in 2013, enjoys the view of Earth from the windows in the ESA built Cupola of the International Space Station. A blue and white part of Earth is visible through some of the seven windows of the Cupola (Photo: NASA)

After long stays in weightlessness, changes are observed in blood composition that can be problematic. Firstly, the number of red blood cells and the haemoglobin level decrease. Decrease of 20–30 percent in number and of 10–15 percent in mass of red blood cells are measured after one to two months in weightlessness.

Secondly, red blood cells of unequal sizes and of abnormal shapes have been also discovered. After six months in microgravity in orbit, up to two percent of ovalized red blood cells have been observed in Russian cosmonauts.

Thirdly, the immune defence system of astronauts diminishes in microgravity after approximatively seven days of flight. One observes a reduction of production of lymphocyte T cells (the white blood cells) that intervene in the immune responses and in antibodies production. This observation did not find so far, a satisfactory fundamental explanation and this problem could be the one that would impede mankind to adapt to long duration space travels in microgravity. Astronauts are more prone to infections in space and they need more time to recover after an infection on ground after their return. The immune system is back to its normal pre-flight level after a period of five to ten days after return to Earth. The interest for this kind of research is actually very high. Indeed, understanding the defence mechanisms of the human organism and the role played by gravity, or its absence, can shed light on fundamental properties of the immune system. These are of importance in fighting certain viral infections, not fully understood yet on Earth, like e.g. human immunodeficiency virus infection and acquired immune deficiency syndrome (HIV/AIDS).

Once more, one notes the principle of transposing in space in microgravity the investigation of a phenomenon existing on Earth, to analyse its fundamental processes and to understand its functioning without gravity, which can often bring a solution to the problem encountered on Earth.

4 Musculo-Skeletal System

4.1 Spine

In microgravity, the first effect that is noticed is the spine extension up to a point that astronauts can gain a few centimetres in height. This is due to the partial decompression of intervertebral discs that do not have to support the weight of the upper body anymore, similar to what we can experience every morning while waking up. On Earth, the lying down position during sleep favours also the unloading of intervertebral discs: we are taller on average of one centimetre in the morning than in the evening. Back on Earth, after flight, this effect disappears of course and height becomes normal again, but with sometime, the risk of having one of the other nerve blocked between discs and vertebrae.

Let us signal also this anecdote of an American astronaut, tall in height but still within the selection requirement for maximum height (1.90 m), and who found himself nonstandard during the duration of a Spacelab mission because of this effect of intervertebral discs unloading.

Furthermore, some astronauts complained of backpains during or after a space flight, probably caused by this phenomenon of spine extension.

4.2 Muscular System

The muscular system atrophy is a second consequence, observed after some days in weightlessness. In particular, the most affected muscles are those that control posture and that contribute to support the body weight on Earth. In microgravity, the natural position that astronauts take is a curved position with legs slightly bent. Indeed, to move in microgravity, you do not need to stand up straight and walk. One floats freely and moves by pushing oneself against a wall, using the action-reaction principle. One notices thus an atrophy, a loss of mass of muscles and the elimination of muscular proteins (see Fig. 38), yielding their catabolism (or degradation). This elimination is associated with a loss of potassium and nitrogenous components in muscles.

The counter-measures that can be applied to avoid these debilitating effects can be summarized in two words: physical exercise. One requests astronauts to sport in orbit. Among the most used means, they are asked to run on small treadmills

Fig. 38 British ESA astronaut Tim Peake operates the Muscle Atrophy Research and Exercise System (MARES) equipment inside the Columbus module. MARES is an ESA facility used for research on musculoskeletal, biomechanical, and neuromuscular human physiology to better understand the effects of microgravity on the muscular system (Photo: NASA/ESA)

(see Fig. 39), to use squat machine or to cycle on ergometer bicycle (which is a bit paradoxical when one realizes that they are already moving at a speed of 28 000 km/h in orbit!).

Fig. 39 ESA astronaut Alexander Gerst tests a running treadmill during parabolic flights on Airbus A300 under supervision of the author. Note the vertical position of the running surface. The entire treadmill setup was installed vertically due to the large vertical dimension of the treadmill and its subsystems, impeding its installation in a horizontal configuration. The subject ran toward the cabin ceiling in 0 g (there is no up or down in 0 g …). He is lying on an elevated mat (in red) during the high g pull-up and pull-out phases (Photo: ESA)

By practicing regularly (more than two hours per day!) and by applying sometime treatments of muscular fibre electrostimulation, astronauts and cosmonauts have no difficulties to readapt upon return to Earth after a more than six months mission.

Finally, bone demineralisation is the most important phenomenon and is addressed in the next section.

12 Bone Demineralisation

The problem of bone demineralisation and the loss of calcium is much more serious and still not completely understood. Still unresolved, this problem could be the second problem that could thwart the hopes of mankind to adapt to space travels in weightlessness.

One observes a demineralization, mainly decalcification, of the bone structure in astronauts, i.e. a loss of calcium and phosphorus (the hydroxyapatite). Why? We still do not know exactly. One could say very approximately that the skeletal bones do not have to fight against gravity and do not have to support the body weight.

Consequently, bones "atrophy" one way or the other and bone cells do not regenerate like on Earth.

During Skylab mission, at the beginning of the 1970s, one measured that, on average, approximately 100 mg of calcium are lost per day. Now, an adult organism contains about one kilogramme of calcium. Imagining that the decalcification process varies linearly with time (which is not the case!), all the calcium of the organism would be lost in less than 30 years, which seems long. But, in reality, much less time would be needed for bones to fragilize to a point where an astronaut, after staying a few months or years in weightlessness in orbit or travelling to another planet, could not land without suffering multiple fractures!

What we know of the bone decalcification problem does not allow us to reach significant conclusions, as several theories oppose each other. None of them answer all the questions, although each of them can explain a part of the problem. One knows that decalcification is related to an atrophy of bone fibrous cells containing calcium, corresponding to the part of the bone that allows the marrow to pass. This effect seems to be irreversible once it has started. The rate of calcium loss varies from an astronaut to another and varies also from a type of bone to another. One can say though, in first approximation, that it depends on the duration of stay in microgravity, but that it starts to appear only after one to two months passed in orbit. Some bones, like the ulna and radius of the arm, have shown an absence of demineralisation and of calcium loss, while other bones containing marrow exhibit an important demineralization. The most important demineralisations have been observed in a Russian cosmonaut, after a six months flight on board the first Russian space station *Salyut*, and are of the order of 8 percent in the heel bone and up to 10 percent in the lumbar vertebrae. On the other hand, results of experiments conducted on Russian cosmonauts during several months flights have shown that bone demineralisation is mainly marked in bones of the lower body part, while some bones of the upper body (like the skull or shoulders) re-mineralised or even gained in calcium. This very surprising result is presently still only partially explained.

This problem of bone decalcification resembles by certain aspects to an illness known on Earth, osteoporosis, affecting mainly elderly people and women after menopause. This sickness yields a change in the structure (demineralization) of bones, associated to losses of calcium, but also of phosphorus, nitrogen and hydrox-yproline protein (a bone component), but the composition stays globally the same. The bone loses in thickness, fragilizes and fractures more easily. By other aspects, demineralisation observed in microgravity is also similar to another sickness, osteo-malacia, for which one observes only a change in the bone composition, only some minerals are eliminated. One sees in this case a calcium reduction with respect to the bone mass, calcium salts do not deposit anymore in the regenerated bone tissues, the bone fragilizes, can fracture or wrongly form.

Numerous experiments conducted in microgravity on astronauts and cosmonauts and on animals (what is called the animal model) yield results, sometime diverging. On one side, one observes an increase of activity of osteoclastic cells, whose role is to eliminate and resorb elements of bone tissues. On the other side, some results show that bone demineralisation would be due to a decrease of activity of osteoblastic cells, responsible for regenerating bone tissues. Is demineralisation due to a decreased process of bone tissue formation or to an increased process of bone tissue resorption, or to a combination of both processes? One must stay cautious and these results need to be confirmed by many more experiments, including at bone cell biological level.

What to think also of the re-calcification of upper body bones, reported by Russian doctors? The first correlation that comes to mind is to associate this observation with body liquid transfer from body lower part to upper part. One has to consider also two categories of bones: the weight bearing bones in normal gravity on Earth (legs, spine, …) and the non-bearing bones (upper part of skeleton). These two categories of bones react differently during the prolonged exposure to microgravity. But by what process would calcium additionally fix on bones?

Bone decalcification yield also secondary effects. Calcium is transported by blood in the organism and eliminated by the kidneys in urine. This transport can yield important calcium deposits in other parts of the human body, e.g. in the cardiovascular system or on soft tissues, with the potential to affect other physiological functions. Decalcification can also cause kidney failure, as kidneys are too solicited to filter calcium. An increase of 50% of calcium has been measures in urine samples of some astronauts on previous missions.

Despite the incomplete knowledge that doctors have of this problem, counter-measures have been devised to allow astronauts to continue to function in orbit without too much problems and to provide them with a faster post-flight rehabilitation on ground. Simple in their principle, these counter-measures are however insufficient to fight bone demineralisation in weightlessness.

Firstly, if calcium is eliminated from the organism, it should be replaced! So, astronauts receive calcium enriched nutrition. While it is commonly found in dairy products on Earth, space menus settle more prosaically for calcium enriched food. Secondly, if gravity absence yields less work for the bones that should bear body weight, they should be put at work! For long duration missions, during several hours a day, astronauts have to wear what is called a "penguin suit" that puts bones in compression and oblige them to "work" against this compression during physical exercises. This "penguin suit" resembles to a sport tracksuit with elastic braces too small for the astronaut size, going from feet to waist and from waist to shoulders and that compress the body from foot to head. This "penguin suit" is now replaced by a set of braces and straps that astronauts have to wear when running on a treadmill (see Fig. 40). The effect is similar but it allows to be used also in a more dynamic environment of running.

Unfortunately, these two counter-measures are not sufficient. The calcium enriched diet slows down calcium losses, but does not stop it. And, wearing "penguin suit" or braces and straps to run are not the astronaut favourite pastime, nor is the fact that they have to spend several hours per day on treadmills, squat apparatuses or ergometric bicycles.

Space medical research on physical training of astronauts has a spin-off in every-day life on Earth for prevention of fractures due to osteoporosis in women after menopause. Appendix 13 presents this simple approach but having important consequences for public health.

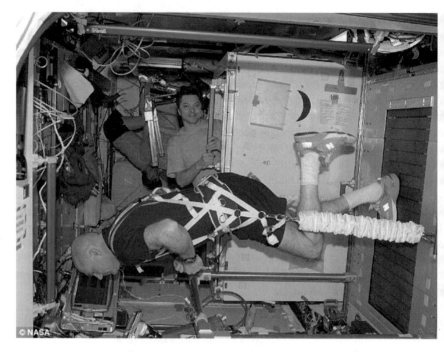

Fig. 40 On ISS, astronauts must spend several hours per day exercising in order to stay in shape. Dutch ESA astronaut Andre Kuipers (foreground) runs on the Combined Operational Load Bearing External Resistance Treadmill (COLBERT) in the Tranquillity Node of ISS in 2012. In the background, another astronaut is seen "upside down" (but there is no up, no down in microgravity…) using a squat apparatus (Photo: NASA)

Conclusions

All, but two or possibly three, physiological modifications observed in astronauts and cosmonauts after a more or less long exposure to microgravity are most likely not impeding the continuation of orbital space flights and eventually to another planet. The first problem of decreased immune defence system after a more or less long stay in weightlessness is an important problem as it concerns the human organism ability to defend itself against bacterial and other infectious agents' aggressions. One noticed as well that some types of bacteria are developing differently in weightlessness, becoming more virulent, or even mutating. With lower immune defences, the combination of these two phenomena present an important challenge for future space missions of very long duration, like Mars missions. The second problem of bone demineralisation is also not solved completely. One manages to prevent the problem in orbit such that astronauts' health is not endangered, as missions of more than six months and the few missions of more than a year have shown. However, fundamental answers to this problem are not yet available. The third problem, the visual impairment and intracranial pressure, may also affect space travellers in long duration interplanetary journeys and is presently actively investigated on the ISS.

One does not know yet the long-term effects of a prolonged microgravity exposure and their consequences on astronauts' mortality. One cannot conclude also whether the human organism could totally adapt to microgravity.

Research and experiments of course continue, but every answer poses new questions.

Space medical research contributed to increase our knowledge on certain medical afflictions and allowed to find treatments that are applied today to patients on Earth. Medical techniques developed for space flights have also spin-offs in the actual medical means used daily in hospitals.

The diversity of physiological modifications observed on the human organism show the wealth of the investigation field offered to researchers involved with problems of adaptation of human beings in space. This also holds true for animal, vegetal and biological organisms. Numerous investigations are conducted in other life science domains like biology, genetics, … and were not addressed here.

Finally, microgravity research allows to come close to an old dream of mankind: to escape from gravity to which we are all constrained on Earth.

Appendix 13: An Application of Space Medicine Research

A spin-off of research on prevention of astronaut bone demineralisation consists in the preparation of a programme of simple physical exercises that is prescribed to women after menopause to avoid fractures due to osteoporosis. Statistically, a woman out of ten will suffer during her life from a femur neck fracture. This frequency is doubled after menopause. After the age of 60, 20% of female patients will decease and half of surviving will be handicapped. It is therefore of paramount importance to understand the functioning mechanisms of osteoporosis and bone fragilization.

On the other hand, preventing consequences of fractures due to osteoporosis, if possible at all, has an obvious economical character, as post-menopause fractural osteoporosis is an important and costly public health problem.

Weightlessness is creating a fast evolution model of bone loss, space research in microgravity on bone pathology can contribute to better understand fundamental mechanisms of osteoporosis. Programmes of preventive physical activity based on acute osteoporosis as observed in microgravity are derived from astronaut training programmes.

So, some French specialists recommend the following types of exercises (always under medical supervision!). Exercises with vertebral extension are recommended, those with flexion are to be avoided. Exercises should be done while under load (walking, running, stair climbing) and should not be necessarily done every day. A programme of moderated exercises (jogging, exercises while standing, lying down or sitting during an hour, twice a week) would already yield an increase of minimal content of vertebral bone. An hour of walking per day during two years would reduce by half the frequency of osteoporotic fractures.

Thus, one very logically recommends to sport to stay in good health!

References

Classical Books on Physical Mechanics

1. Arnold, V.I.: Mathematical Methods of Classical Mechanics. Springer, New York (1978)
2. Brouwer, D., Clemence, G.M.: Methods of Celestial Mechanics. Academic Press, New York (1961)
3. Goldstein, H.: Classical Mechanics, 2nd edn. Addison Wesley (1980)
4. Landau, L.D., Lifshitz, E.M.: Mechanics. In: Course of Theoretical Physics, vol. 1, 2nd edn. Pergamon Press, Oxford (1969)
5. Roy, A.E.: Orbital Motion, 3rd edn. Adam Hilger, Bristol (1988)
6. Spiegel, M.R.: Theoretical Mechanics. Schaum's Outline Series in Sciences. McGraw-Hill, New York (1967)
7. Vallado, D.A.: Fundamentals of Astrodynamics and Applications. McGraw-Hill Companies, Inc. (1997)

Books on Microgravity Research

8. Beysens, D.A., Van Loon, J., et al. (eds.) Generate an Extra-Terrestrial Environment on Earth. River Publishers Series in Standardisation, Aalborg, Denmark, ISBN 978-87-93237-53-7 (2015)
9. Seibert, G., et al.: A world without gravity. In: Fitton, B., Battrick, B. (eds.): ESA-SP-1251, ESA-ESTEC, ISBN 92-9092-604-X (2001). http://www.esa.int/esapub/sp/sp1251/sp1251web.pdf. Last accessed 31 July 2017
10. Haerendel, G., et al.: Looking Up, Europe's Quiet Revolution in Microgravity Research. Scientific American, custom publication (2008). https://www.scientificamerican.com/media/pdf/ESAReader_LowRes.pdf. Last accessed 31 July 2017
11. Myshkis, A.D., Babskii, V.G., Kopachevskii, N.D., Slobozhanin, L.A., Tyuptsov, A.D.: Low-Gravity Fluid Mechanics. Mathematical Theory of Capillary Phenomena. Springer, Berlin (1987)
12. Clément, G.: Fundamentals of Space Medicine. Springer, New York (2005)
13. Clément, G., Reschke M.F.: Neurosciences in Space. Springer, New York, ISBN 9780387789507-9780387789491 (2008). https://doi.org/10.1007/978-0-387-78950-7

© The Author(s) 2018
V. Pletser, *Gravity, Weight and Their Absence*, SpringerBriefs in Physics,
https://doi.org/10.1007/978-981-10-8696-0

On Drop Towers

14. The ZARM Drop Tower. https://www.zarm.uni-bremen.de/drop-tower.html. Last accessed 31 July 2017
15. Von Kampen, P., Kaczmarczik, U., Rath, H.J.: The new drop tower catapult system. Acta Astronaut. **59**, 278–283 (2006)
16. Wei, M.G., Tian, Q.L., Chi, Z.H., Wan, S.X., Hu, W.R.: Recent progress in NMLC drop tower. In: Proceedings of 2nd China-Germany Workshop on Microgravity Science, Beijing, China, vol. 2, p. 263 (2002)
17. Liu, T.Y., Wu, Q.P., Sun, B.Q., Han, F.T.: Microgravity level measurement of the Beijing drop tower using a sensitive accelerometer. Sci. Rep. **6**, Article no. 31632 (2016). https://doi.org/10.1038/srep31632
18. Introduction to Key Laboratory of Microgravity, Institute of Mechanics, Chinese Academy of Sciences. http://english.imech.cas.cn/rh/rd/nml/. Last accessed 31 July 2017
19. Melville, N.: Drop tower. In: Sabbatini, M., Sentse, N. (eds.) ESA User Guide to Low Gravity Platforms, HSO-K/MS/01/14, Iss. 3 Rev. 0, pp. 4-1–4-19 (2014). http://wsn.spaceflight.esa.int/docs/EUG2LGPr3/EUG2LGPr3-4-DropTower.pdf

On Parabolic Flights

20. Pletser, V., Kumei, Y.: Parabolic flights. In: Beysens, D.A., Van Loon, J. (eds.) Generate an Extra-Terrestrial Environment on Earth. River Publishers Series in Standardisation, Aalborg, Denmark, ISBN 978-87-93237-53-7, Chap 7, pp. 61–74 (2015)
21. Pletser, V.: European aircraft parabolic flights for microgravity research, applications and exploration: a review. Review paper, REACH—Reviews in Human Space Exploration, vol. 1, pp. 11–19 (2016). http://www.sciencedirect.com/science/article/pii/S2352309316300049
22. Pletser, V., Rouquette, S., Friedrich, U., Clervoy, J.F., Gharib, T., Gai, F., Mora, C.: European parabolic flight campaigns with Airbus A300 ZERO-G: looking back at the A300 and looking forward to the A310. Adv. Space Res., 1003–1013 (2015)
23. Pletser, V., Rouquette, S., Friedrich, U., Clervoy, J.F., Gharib, T., Gai, F., Mora, C.: The first European parabolic flight campaign with the Airbus A310 ZERO-G. Microgravity Sci. Technol. **28**, 587 (2016). https://link.springer.com/article/10.1007/s12217-016-9515-8. http://rdcu.be/jRcH
24. Pletser, V., Winter, J., Duclos, F., Friedrich, U., Clervoy, J.F., Gharib, T., Gai, F., Mora, C.: The first joint European Partial-g parabolic flight campaign at Moon and Mars gravity levels for science and exploration. Microgravity Sci. Technol. (special Issue ELGRA 2011) **24**(6), 383–395 (2012). https://link.springer.com/article/10.1007/s12217-012-9304-y
25. Pletser, V.: Are aircraft parabolic flights really parabolic? Acta Astronaut. **89**, 226–228 (2013). https://doi.org/10.1016/j.actaastro.2013.04.019
26. Pletser, V., Harrod, J.: The Science of Gravity; A new era of ESA experiments on parabolic flights. ESA Bull **160**, 22–33 (2014). http://esamultimedia.esa.int/multimedia/publications/ESABulletin160/5
27. Zero-G Experiment Center (OGEC): CSU-CAS-June (2017). http://www.zerogec.ac.cn/. Last accessed 31 July 2017
28. Pletser, V. Parabolic flights. In: Sabbatini, M., Sentse, N. (eds.) ESA User Guide to Low Gravity Platforms, HSO-K/MS/01/14, Iss. 3 Rev. 0, pp. 5-1–5-22 (2014). http://wsn.spaceflight.esa.int/docs/EUG2LGPr3/EUG2LGPr3-5-ParabolicFlights.pdf

On Sounding Rockets

29. Mawn, S.: REXUS User Manual, ref. RX_UserManual_v7-13_15Dec14.doc, EuroLaunch (2014)
30. Verga, A.: Sounding rockets. In: Sabbatini, M., Sentse, N. (eds.) ESA User Guide to Low Gravity Platforms, HSO-K/MS/01/14, Iss. 3 Rev. 0, pp. 6-1–6-31 (2014). http://wsn.spaceflight.esa.int/docs/EUG2LGPr3/EUG2LGPr3-6-SoundingRockets.pdf

On International Space Station

31. Nasa website. https://www.nasa.gov/mission_pages/station/main/
32. ESA websites. http://www.esa.int/Our_Activities/Human_Spaceflight/International_Space_Station/Highlights/International_Space_Station_panoramic_tour, http://www.esa.int/Our_Activities/Human_Spaceflight/International_Space_Station/Building_the_International_Space_Station3
33. Sabbatini, M., Sentse, N. (eds.): International Space Station—ISS: ESA User Guide to Low Gravity Platforms, HSO-K/MS/01/14, Iss. 3 Rev. 0, pp. 7-1–7-106 (2014). http://wsn.spaceflight.esa.int/docs/EUG2LGPr3/EUG2LGPr3-7-ISS.pdf

On Tiangong-2 Mission

34. Yang, Y., Chen, M., Cao, Q., Pletser, V.: Gorgeous Sky Palace, Laboratory Above the Sky. (in Chinese), Technology and Engineering Centre for Space Utilization CSU, ScienceP Publ., Beijing, ISBN 978-7-03-049870-0 (2016)

On Appendix 12, an Example of Application of Microgravity Research

35. Ratke, L. (ed.): Immiscible Liquid Metals and Organics. DGM Informationsgesellschaft-Verlag (1993)
36. Walter, H.U.: Microgravity Research Spin off and Applications-Example: Marangoni Convection. Microgravity News from ESA, 6-3, 2-4, ESA Paris (1993)

On Physiological Problems in Microgravity

37. Aubert, A., Beckers, F., Verheyden, B., Pletser, V.: What Happens to the Human Heart in Space?—Parabolic Flights Provide Some Answers. Cardiovascular response during gravity changes induced by parabolic flights. ESA Bull. 119, 30–38 (2004). http://www.esa.int/esapub/bulletin/bulletin119/bul119_chap4.pdf
38. Preidt, R.: Spaceflight Might Weaken Astronauts' Immune Systems. HealthDay, 30 Aug. 2014. https://consumer.healthday.com/kids-health-information-23/immunization-news-405/spaceflight-might-weaken-astronauts-immune-systems-690922.html

39. American Society for Biochemistry and Molecular Biology (ASBMB): The human immune system in space. ScienceDaily, 22 April 2013. www.sciencedaily.com/releases/2013/04/ 130422132504.htm
40. Sutton, J.: How does spending prolonged time in microgravity affect the bodies of astronauts? Sci. Am. (2003). https://www.scientificamerican.com/article/how-does-spending-prolong/

Printed in the United States
By Bookmasters